SOLID STATE CHEMICAL SENSORS

Contributors

Robert J. Huber
Jiří Janata
Ingemar Lundström
Christer Svensson
Jay N. Zemel

SOLID STATE CHEMICAL SENSORS

Edited by JIŘÍ JANATA

Department of Bioengineering
The University of Utah
Salt Lake City, Utah

ROBERT J. HUBER

Department of Electrical Engineering
The University of Utah
Salt Lake City, Utah

 1985

ACADEMIC PRESS, INC.

(Harcourt Brace Jovanovich, Publishers)

Orlando   San Diego   New York   London

Toronto   Montreal   Sydney   Tokyo

ACADEMIC PRESS, INC.
Orlando, Florida 32887

United Kingdom Edition published by
ACADEMIC PRESS INC. (LONDON) LTD.
24–28 Oval Road, London NW1 7DX

Library of Congress Cataloging in Publication Data
Main entry under title:

Solid state chemical sensors.

   Includes index.
   1. Chemical detectors. 2. Solid state chemistry.
I. Janata, Jiří. II. Huber, Robert J.
TP159.C46S65 1985   681'.2   84-20374
ISBN 0–12–380210–5 (alk. paper)

PRINTED IN THE UNITED STATES OF AMERICA

85 86 87 88   9 8 7 6 5 4 3 2 1

Contents

4.   An Introduction to Piezoelectric
       and Pyroelectric Chemical Sensors

       Jay N. Zemel

List of Contributors

Numbers in parentheses indicate the pages on which the authors' contributions begin.

ROBERT J. HUBER (119), Department of Electrical Engineering, The University of Utah, Salt Lake City, Utah 84112

JIŘÍ JANATA (65), Department of Bioengineering, The University of Utah, Salt Lake City, Utah 84112

INGEMAR LUNDSTRÖM (1), Laboratory of Applied Physics, Department of Physics and Measurement Technology, Linköping Institute of Technology, S-581 83 Linköping, Sweden

CHRISTER SVENSSON (1), Laboratory of Applied Physics, Department of Physics and Measurement Technology, Linköping Institute of Technology, S-581 83 Linköping, Sweden

JAY N. ZEMEL (163), Center for Chemical Electronics, Department of Electrical Engineering, University of Pennsylvania, Philadelphia, Pennsylvania 19104

Preface

Detection of chemical species using solid state circuitry is a relatively new field that is generating a great deal of interest, both in academia and in industry. Several conference proceedings have been published on this subject, and review articles have been written that deal with particular subsets of solid state sensors. This volume is our attempt to review the basic chemical and physical principles—and problems—involved in the construction and operation of some of these devices.

A major portion of the book is devoted to explanation of the basic mechanism of operation and the many actual and potential applications of field effect transistors for gas and solution sensing. A chapter describing the basics of device fabrication is included so that the nonspecialist reader may gain an appreciation of the complexity of semiconductor fabrication methods. The chapter on piezoelectric and pyroelectric chemical sensors outlines early work in the development of new techniques of chemical detection. Chemical sensing covers a vast territory, and it was necessary to omit many areas of important research, e.g., high-temperature surface conductivity sensors, chemiresistors, etc. Although this volume is not intended to be used as a textbook, some of the material is suitable for inclusion in graduate courses.

This emerging technology is dependent on the research efforts of two groups of scientists, electrical engineers and chemists, because solid state chemical sensors are hybrid devices that employ the principles of both these fields. There are many similarities in the laws that describe the seemingly quite different phenomena in these two disciplines. Our major goal in this book, then, is to demonstrate this coincidence in the expressions of these phenomena and to use it to assist electrical engineers in understanding the chemistry involved and to educate chemists in solid state science. We hope that our efforts will help accelerate progress in the exciting new field of solid state chemical sensors.

1

Gas-Sensitive Metal Gate Semiconductor Devices

INGEMAR LUNDSTRÖM AND
CHRISTER SVENSSON

LABORATORY OF APPLIED PHYSICS
DEPARTMENT OF PHYSICS AND MEASUREMENT TECHNOLOGY
LINKÖPING INSTITUTE OF TECHNOLOGY
LINKÖPING, SWEDEN

I. Introduction

The first descriptions of a hydrogen-sensitive metal–oxide–semiconductor (MOS) field effect transistor were published in 1975 (Lundström *et al.*, 1975a,b). This device represents—to our knowledge—the first application of a chemically active metal gate, namely palladium, in an active semiconductor device. Since then a number of papers have been published on the subject. Most studies on the Pd-MOS devices have been made in Sweden, but similar devices have also been studied by other groups, as reviewed in the following. This chapter deals mainly with Pd–SiO_2–Si structures, although other devices, such as Pd–semiconductor Schottky barriers, are also considered. Our main purpose is to give a simple physical description of semiconductor devices with catalytic metal gates. In addition, we indicate the present level of understanding of these devices, their drawbacks, and their promise. Special attention is paid to the behavior of hydrogen in the Pd–SiO_2 system. Not only the wanted signal but also some hysteresis and long-term drift phenomena are due to the properties of this system. A description of some applications of hydrogen-sensitive transistors is given, e.g., smoke detection and biochemical reaction monitoring.

II. MOS Device Physics

A. INTRODUCTION TO THE SEMICONDUCTOR SURFACE

Semiconductors in general contain relatively few free charge carriers. This facilitates control of the concentration and behavior of these charge carriers by external means. Many semiconductor phenomena have therefore became very attractive for technical applications. Furthermore, one semiconductor, silicon, has excellent stability, as has its oxide, silicon dioxide. The silicon–silicon dioxide system has therefore made semiconductor technology perfectly suited for industrial products. This technology has already led to an industrial revolution in electronics and information science, and it is natural to seek even more applications for it—for example, in chemical sensors.

The most important property of a semiconductor is its concentration of charge carriers. In an absolutely clean semiconductor there are equal amounts of negative free electrons and positive free holes, created by the excitation of valence electrons from the valance states in the crystal to the first band of excited states, called the conduction band (see Fig. 1). By such thermal excitation free electrons are formed in the conduction band, leaving unoccupied valence states. These unoccupied states are also considered as free charge carriers, positively charged (as the crystal was neutral when the state was filled), and are called holes. The concentration of these two carriers is controlled by the mass action law applied to the reaction $h^+ + e^- \rightleftarrows$ crystal:

$$np = n_i^2 \tag{1}$$

where n and p are the electron and hole concentrations, respectively, and n_i is a constant. For the clean semiconductor $n = p$, because it is electrically neutral; thus both n and p are equal to n_i, the intrinsic carrier concentration.

The intrinsic carrier concentration is very small in silicon, about 10^{16} m^{-3} at room temperature. However, n or p may be increased by doping. Consider a semiconductor doped by N_D positive ions per unit volume: the ions are assumed completely dissociated. We then have two equations to fulfill, Eq. (1) and the electrical neutrality condition

$$n = p + N_D \tag{2}$$

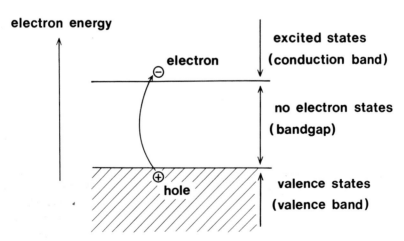

Fig. 1. Energy bands characterizing a semiconductor. Free electrons occur at the bottom of the conduction band. A hole corresponds to the lack of a valence electron and behaves like a mobile positive charge.

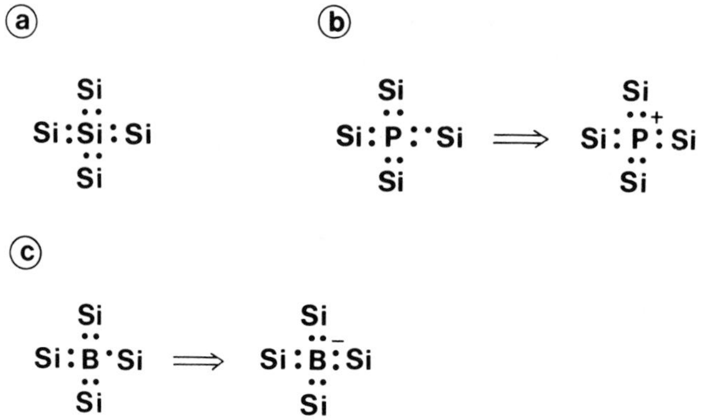

(a)

```
        Si
        ••
    Si:Si:Si
        ••
        Si
```

(b)

```
        Si                          Si
        ••                          ••  +
    Si:P:•Si    ⟹            Si:P:Si
        ••                          ••
        Si                          Si
```

(c)

```
        Si                          Si
        ••                          ••  −
    Si:B•Si    ⟹            Si:B:Si
        ••                          ••
        Si                          Si
```

Fig. 2. Chemical bonding structure of silicon (a) compared to silicon doped with phosphorus (b) and boron (c). In the case of boron one electron is taken from the surrounding silicon atoms (giving rise to a hole) to complete the sp^3 hybrid bonds. In the case of phosphorus one electron does not fit into the sp^3 bonds and is therefore free (donated to the conduction band).

Normally, N_D is chosen much larger than n_i; N_D determines n, thus $n = N_D$. We have formed an n-type semiconductor, dominated by electron conduction. Note that the ions are fixed in the crystal. In silicon these ions may be, for example, phosphorus, which forms ions at silicon lattice positions, as shown in Fig. 2b. In the same way we may form a p-type semiconductor by doping with negatively charged ions, for example, boron (Fig. 2c).

The electron concentration can also be described in terms of the chemical potential or Fermi energy ϕ_F

$$n = n_i \exp\left(\frac{q\phi_F}{kT}\right) \tag{3a}$$

where q is the electron charge. As holes are just lack of electrons they are related to the same chemical potential or Fermi energy; thus from Eq. (1):

$$p = n_i \exp\left(\frac{-q\phi_F}{kT}\right) \tag{3b}$$

Energy bands and the Fermi energy are often represented in a "band diagram," shown in Fig. 3 as electron energy versus some space parameter (x axis). The band of valence states, the valence band, is shown normally filled with electrons in a simple covalent material. Holes will occur at the top of the valence band. The band of conduction states (first excited states), the conduction band, is normally empty. Free electrons occur at the bottom of this band. The Fermi energy is represented in this

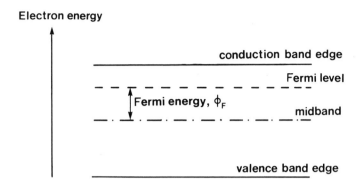

Fig. 3. Simplified energy band diagram for a semiconductor. The Fermi level and the Fermi energy are defined. (The example is for an n-type semiconductor.)

diagram as a dotted line. It is easy to remember that more electrons occur if this line is close to the conduction band (making ϕ_F large), and vice versa for holes.

Let us now consider a semiconductor surface. Figure 4a shows a band diagram of a p-type semiconductor with a surface that does not disturb the interior of the semiconductor. We have total electrical neutrality and constant potentials. Assuming that the surface contains a positive charge,

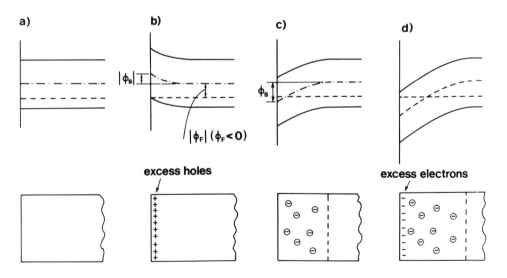

Fig. 4. Schematic illustration of the possible charge states of a p-type semiconductor and the corresponding energy band diagrams: (a) neutral, (b) accumulation, (c) depletion, (d) inversion.

the band diagram will be as shown in Fig. 4b. This situation occurs, for example, if we have a negative potential or negative charge somewhere outside the semiconductor surface. The positive charge in the surface consists of an extra hole concentration. Because the surface is no longer neutral, there is also a variable electric potential ϕ. In Fig. 4 this potential is shown as a corresponding change in electron energy, that is, as a bending of the conduction and valence band edges. The hole concentration is, of course, associated with the potential (note that $\phi_F < 0$ in a p-type material):

$$p = n_i \exp \frac{q}{kT}(-\phi - \phi_F) \tag{4a}$$

where ϕ becomes more negative closer to the surface. The change in semiconductor potential is largest at the surface. The value of the potential at the surface is called the surface potential ϕ_S. The situation discussed is called the accumulation regime, because holes accumulate at the surface.

In the case of a positive external charge or potential the semiconductor becomes negatively charged. The first effect of a positive external charge is to push the free holes away from the surface, leaving negative ions behind (Fig. 4c). Again, the hole concentration follows Eq. (4a). As ϕ is now positive the hole concentration is very small and we have a region in the semiconductor that is depleted of holes, the depletion region. The width of this region depends on doping, that is, on the concentration of negative ions, and is of the order of 0.1–1 μm in most cases. In contrast, the accumulation region, described above, is much narrower. The situation depicted in Fig. 4c is called the depletion regime. Note also that the electron concentration is controlled by an equation similar to Eq. (4a):

$$n = n_i \exp \frac{q}{kT}(\phi + \phi_F) \tag{4b}$$

Thus, if the potential ϕ changes enough the electron concentration may be of importance. Specifically, for $\phi_S = 2|\phi_F|$, the electron concentration at the surface is equal to the bulk hole concentration. This is the point of onset of the next regime, the inversion regime, which occurs for a higher external positive potential or charge. The inversion regime is demonstrated in Fig. 4d. The surface has been inverted from p-type to n-type, and a channel of electrons has formed at the surface. This channel is the basis of the MOS transistor discussed below. After reaching the inversion regime, all new charges are accumulated in the channel; thus the depletion width remains constant.

B. MOS Devices

MOS devices are normally based on the semiconductor silicon. If single-crystal silicon is oxidized in oxygen or water vapor at a high temperature, a high-quality silicon dioxide film is formed on the surface. The discovery of this phenomenon is the basis of the evolution of modern integrated circuit technology and the MOS technology. An MOS device is one based on the combination metal–oxide–silicon.

MOS devices can be made in two forms, as MOS capacitors or MOS transistors (Fig. 5). The devices can be made out of p-type silicon doped with trivalent impurities, such as boron. The impurities exist as negative ions in the crystal, and they will be compensated by an equal concentration of free holes (positive). The two concentrations are of the order of 10^{21} m^{-3}. The p-type silicon wafer is oxidized to an oxide thickness of the order of 100 nm and covered by a metal dot—for example, aluminum of about the same thickness. In the case of the transistor we have also formed two n-doped contacts in the silicon crystal. These areas are doped with, for example, phosphorus, and contain free electrons. The distance between the two regions is of the order of 10 μm.

For a negative potential on the metal electrode (or gate) the silicon surface will be charged positively through a higher concentration of holes. The surface will have metallic conductivity because of the high concentration of free charges, and the MOS capacitor will act like a normal plate capacitor with the oxide as a dielectric, having a capacitance equal to C_0 (Fig. 6). A positive potential on the gate pushes the holes away from the silicon surface, leaving only the negatively charged doping atoms near the surface, and these now form the negative charge on the silicon side of the capacitor. The capacitance of the MOS structure will be lower than before, because it is acting as a plate capacitor with a large dielectric thick-

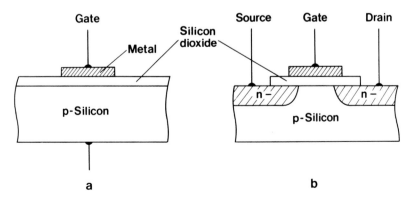

Fig. 5. The basic MOS capacitor (a) and MOS transistor (b).

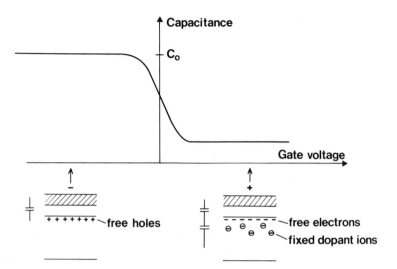

Fig. 6. The capacitance–voltage curve for an MOS capacitor. Schematic pictures of the charge distribution in the capacitor for negative and positive voltages are also shown.

ness: the oxide thickness plus the thickness of the depleted layer in the silicon. If we increase the gate voltage further the capacitance will continue to decrease until it reaches a point at which the electron concentration starts to increase at the silicon surface. In other words, the electric potential in the surface has lowered the hole concentration so much that the electron concentration increases according to the mass action law, forming an inversion layer. Figure 6 shows the total capacitance versus voltage curve that is used for analyzing the MOS capacitor.

In the case of a transistor we observe the conductance between the two n-type electrodes, the source and the drain, when the gate voltage is changed. With no gate voltage applied the conductance is almost zero. The two electrodes act as two opposite pn junctions or diodes, one diode always operating in its reverse direction. This is true in all situations except when an inversion layer is formed. An inversion layer forms an n-type channel between the two n-type contacts, and this channel provides a conductive path between the source and the drain. Figure 7 shows the drain–source conductance versus the gate voltage at small drain voltages. The conductance is zero until we reach the threshold voltage V_T at which the inversion layer starts to form. Then the conductance increases linearly with the gate voltage.

The gate capacitance stores a channel charge Q given by

$$Q = C_g(V_G - V_T) \tag{5}$$

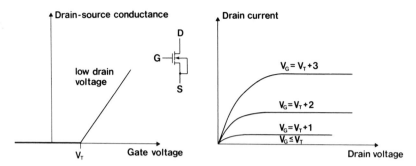

Fig. 7. Characteristic curves for an MOS transistor. The symbol of the *n*-channel MOS transistor is also shown. The arrow indicates the *p*-type substrate, which is normally connected to the source.

where C_g is the gate capacitance (C_0 times gate area), V_g is the gate voltage, and V_T is the gate voltage at which the channel starts to form, called the threshold voltage (Q is actually negative for an *n*-channel device). The gate charge passes the channel in a time t, given by the channel length L divided by the electron velocity v. The velocity is given by the electron mobility μ times the electric field along the channel, V_D/L

$$t = \frac{L}{v} = \frac{L^2}{\mu V_D} \tag{6}$$

From this we can calculate the transistor current I_D, given by the amount of charge that passes the channel per unit time

$$I_D = \frac{Q}{t} = \left(\frac{\mu C_g}{L^2}\right)(V_G - V_T)V_D \tag{7}$$

This formula is, in fact, correct for small drain voltages. At larger drain voltages, the potential varies along the channel and the induced charge $q(x)$ in the channel (per unit length) is a function of distance along the channel

$$q(x) = \frac{C_g}{L}[V_G - V_T - V(x)] \tag{8}$$

Furthermore, the electric field along the channel is $-dV/dx$ and the current is

$$I_D = q(x)\mu \frac{dV}{dx}$$

I_D must be constant along the channel, and by integrating the expression above with $q(x)$ given by Eq. (8) from $x = 0$ to $x = L$ (or from 0 to $V = V_D$)

we find that

$$I_D = \frac{\mu C_g}{L^2}\left[(V_G - V_T)V_D - \frac{V_D^2}{2}\right] \tag{9}$$

We observe that Eq. (7) is a special case of Eq. (9).

For $V_D = V_G - V_T$ the induced mobile charge $q(x)$ becomes zero at the drain; the channel is said to be pinched off. If $V_D > V_G - V_T$, the pinch-off point [where $V(x) = V_G - V_T$] moves away from the drain and a depletion region occurs at the drain contact. Current continuity is sustained through a rapid transfer of the channel charge, which reaches the pinch-off point, across the depletion region. For $V_D > V_G - V_T$, the drain current stays constant and is given by

$$I_D = \frac{\mu C_g}{L^2}\frac{(V_G - V_T)^2}{2} \tag{10}$$

which was obtained by replacing V_D with $V_G - V_T$ in Eq. (9). The region $V_D > V_G - V_T$ is called the saturation region. A small increase in the current is observed when V_D is in the saturation region, because the effective channel length becomes smaller. A more thorough treatment of the transistor current can be found in Sze (1981) and Yang (1978).

The value of the threshold voltage or the flat-band voltage depends on several factors. The flat-band voltage is the voltage at which the silicon surface field is zero (or silicon surface charge is zero). It is given by (Sze, 1981; Yang, 1978)

$$V_{FB} = \phi_m - \chi_s - \frac{E_g}{2} - \phi_F - \frac{W_{ox}}{\varepsilon_{ox}}Q_s \tag{11}$$

where ϕ_m is the metal work function, χ_s is the semiconductor electron affinity, E_g is the silicon band gap, ϕ_F is the difference between the silicon midband and the Fermi level in the bulk silicon, W_{ox} and ε_{ox} are the oxide thickness and the dielectric constant, respectively, and Q_s is the oxide charge. The first four terms represent the difference in work function (or difference in electrochemical potential) between the two materials involved, the metal and the semiconductor. The last term represents the effect of charge at the oxide–semiconductor interface of value Q_s per unit area. This charge may be important for device stability and is discussed in Section VI. The equation is further explained by the band diagram in Fig. 8.

The threshold voltage is given by (Sze, 1981; Yang, 1978)

$$V_T = V_{FB} + 2\phi_F + \frac{W_{ox}}{\varepsilon_{ox}}\sqrt{4q\varepsilon_s N_A \phi_F} \tag{12}$$

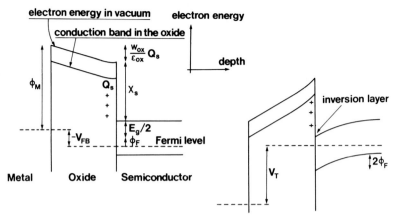

Fig. 8. Energy band diagrams illustrating the concepts of flat-band voltage V_{FB} and threshold voltage V_T. In the examples shown $V_{FB} < 0$ and $V_T > 0$.

where ε_s is the dielectric constant of silicon and N_A is the concentration of doping impurities.

The second term in Eq. (12) represents the difference in interface potential between the flat-band condition and the onset of inversion (onset of a channel). The third term represents the effect of the negatively charged doping atoms at the onset of inversion (see also Fig. 8). Most important here is that V_{FB} and thus V_T depend linearly on the metal work function. This fact is used in MOS gas sensors to detect substances by the change they cause in the metal work function. The effect of all the other parameters on the threshold voltage is of importance for device stability, as some of these parameters will change with time and temperature.

[For further details about MOS devices see the books by Sze (1981) and Yang (1978).]

III.   Palladium Gate MOS Devices

A.   THE GAS-SENSITIVE PD-GATE DEVICES

A gas-sensitive MOS device was first reported by Lundström *et al.* (1975a). This device was a Pd-gate MOS transistor with very thin oxide and Pd layers (about 10 nm). Later, gas sensitivity was observed in transistors and capacitors with both thin and thick oxide and Pd layers (Lundström *et al.*, 1975b, 1977; Steele *et al.*, 1976).

A simple and basic measurement on a Pd-MOS transistor is illustrated in Fig. 9. The transistor is connected as an MOS diode, that is, with the drain and gate electrodes connected together. The MOS diode will then

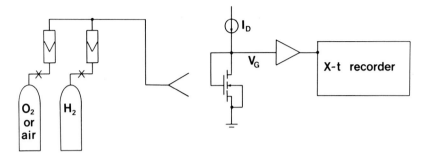

Fig. 9. Experimental setup for measuring hydrogen sensitivity of Pd-gate MOS transistors.

have the characteristic given in Eq. (10). If the MOS diode is connected to a constant-current source as in Fig. 9, the voltage across it will depend linearly on V_T. For a small value of the current the voltage will be nearly equal to V_T. The transistor is heated and placed in a stream of gas of controlled composition. The voltage is recorded as a function of time as the gas composition is changed; the gas composition is controlled by flowmeters, as shown in the drawing.

Figure 10 shows some experimental results, namely the voltage changes with time when the gas composition was changed from air to air with hydrogen and back again. The voltage was thus changed from one stable value to another stable value within about 10 s. The stable values were then recorded versus the hydrogen partial pressure (Fig. 11), with results that fit well an equation of the form

$$\Delta V = \frac{\Delta V_{max} C \sqrt{P_{H_2}}}{1 + C \sqrt{P_{H_2}}} \qquad (13)$$

This equation has the form of a so-called Langmuir isotherm (Bond, 1974), where ΔV_{max} is the maximum observable voltage shift, C is a constant, and P_{H_2} is the partial pressure of hydrogen. The device is sensitive to hydrogen, with a sensitivity that increases with lower oxygen pressures. The sensitivity is also high in air: 1 ppm hydrogen can easily be detected.

The mechanism of the observed hydrogen sensitivity has been identified by Lundström *et al.* (1975a,b, 1977). Molecular hydrogen dissociates to atomic hydrogen on the palladium surface. The atomic hydrogen diffuses into the palladium film and some of it adsorbs at the inner palladium surface. The adsorbed hydrogen atoms act as dipoles at the metal–insulator interface and give rise to a change in the work function of the metal at the interface (see Fig. 12), which affects the threshold voltage of a transis-

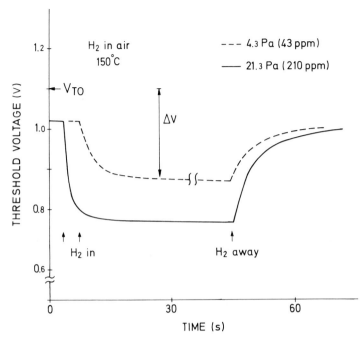

Fig. 10. Effect of hydrogen on the threshold voltage of a Pd-gate MOS transistor. V_{T0} is the threshold voltage for a completely hydrogen-free atmosphere. The background hydrogen in air thus gives a noticeable voltage shift.

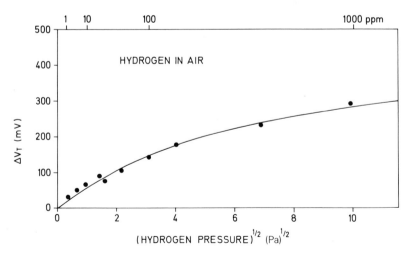

Fig. 11. Change in threshold voltage versus square root of the hydrogen pressure for a Pd-gate MOS transistor. Operating temperature is about 150°C.

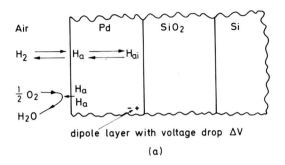

dipole layer with voltage drop ΔV

(a)

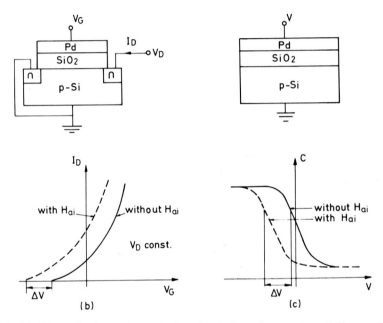

(b)   (c)

Fig. 12. Schematic picture demonstrating chemical reactions on the palladium surface, hydrogen transport, and surface and interface adsorption of atomic hydrogen. (a) The effects of the hydrogen dipole layer at the interface on a Pd-MOS transistor (b) and capacitor (c), respectively, are also indicated.

tor and the flat-band voltage of a capacitor according to Eqs. (11) and (12). Thus, the dipole layer gives rise to an extra voltage in series with the externally applied voltage. The actual change in work function is assumed to be proportional to the interface concentration of adsorbed hydrogen, or the hydrogen coverage. The maximum change occurs when the hydrogen coverage is one, that is, each interface site is occupied by one hydrogen atom. The maximum change ΔV_{max} has been found (by extrapolation) to be about 0.5 V.

The hydrogen coverage at a certain hydrogen pressure depends on two processes. One process is hydrogen dissociation, whereby hydrogen atoms flow into the palladium film. The other process is the way hydrogen flows out again. Hydrogen is lost either through recombination back to molecular hydrogen (in a vacuum or in argon) or through water formation and subsequent water evaporation (in the presence of oxygen). The actual steady-state hydrogen coverage is a result of a balance between these processes. This leads to a considerably higher sensitivity in an inert atmosphere than in, for example, air, as the probability of water formation in the presence of oxygen is much larger than the probability of spontaneous hydrogen recombination. At 150°C a pressure of about 100 Pa in air or 0.01 Pa in argon gives rise to a voltage shift of 0.25 V.

B. Pd-Gate Devices as Hydrogen Sensors

The Pd-gate device can be used as a sensitive detector for hydrogen gas in air or in other environments. It is normally operated at an elevated temperature of 60–150°C in order to decrease the response time and avoid water adsorption, although room-temperature operation is possible.

As shown above, the device gives a fast and reversible response to hydrogen. Its sensitivity depends on the oxygen pressure and sometimes also on the nitrogen pressure. The dependence on oxygen pressure can in many cases be approximated by (Lundström et al., 1975a; Lundström, 1981)

$$\Delta V = \Delta V_{max} \frac{C \sqrt{P_{H_2}/P_{O_2}}}{1 + C \sqrt{P_{H_2}/P_{O_2}}} \tag{14}$$

as demonstrated in Fig. 13. The temperature dependence of the sensitivity is weak in air or oxygen (Lundström et al., 1977). In argon at high temperatures and low hydrogen pressures the sensitivity decreases with temperature, with an activation energy of about 1.1 eV, which corresponds to the heat of adsorption per hydrogen molecule. Equation (14) has been found to be valid for most but not all test structures.

The response time on the introduction of hydrogen is of the order of 5 s (for 50 ppm hydrogen in air at 150°C). It decreases with temperature with an activation energy of about 0.3 eV (Lundström et al., 1977). Furthermore, the response time increases with oxygen pressure, probably due to increased coverage of oxygen on the metal surface (Plihal, 1977; Lundström and Söderberg, 1981–1982).

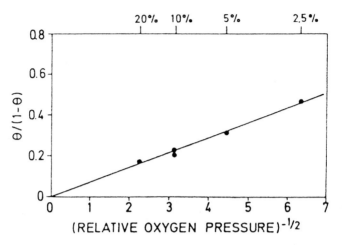

Fig. 13. Example of sensitivity of Pd-gate devices to oxygen pressure. θ is the hydrogen coverage at the interface, $\theta = \Delta V/\Delta V_{max}$. θ is discussed in Section V.

A more detailed description of the metal surface reactions controlling the gas sensitivity and response times is given in Section V.

The devices also show a number of unwanted effects such as drift, hysteresis, and inactivation. These phenomena will be discussed in Section III,E.

C. OTHER GASES

The simple Pd-gate devices have also been observed to be sensitive to hydrogen sulfide (Shivaraman, 1976) and ammonia (Lundström et al., 1975b). In both cases the sensitivity is explained in terms of dissociation of the gas and subsequent diffusion of atomic hydrogen to the metal–silicon dioxide interface, as described above. Table I shows approximate comparisons among sensitivities for the different gases. In the case of hydrogen sulfide no poisoning is observed in air, probably because of the effect of oxygen. In an inert atmosphere, however, the devices are observed to be poisoned by hydrogen sulfide. It should be noted that not all Pd-gate devices respond to ammonia; the difference between a Pd device that responds to ammonia and one that does not probably lies in the structure of the metal film and the properties of the insulator surface (see also Section V).

By replacing the simple palladium gate with other materials the device may be made sensitive to other gases. The use of porous palladium in such a device has been described (Söderberg et al., 1980; Dobos et al., 1980; Yamamoto et al., 1980; Lundström and Söderberg, 1981–1982).

TABLE I

ESTIMATED DETECTION LIMITS OF PALLADIUM
DEVICES (SENSOR TEMPERATURE 150°C)

Gas	Estimated detection limit for minimum detectable voltage shift ΔV_{min}	
	10 mV	1 mV
H_2		
In air	0.5 ppm	0.005 ppm
In inert atmosphere	0.03 ppb	0.0003 ppb
H_2S in air	5 ppm	0.05 ppm
NH_3 in air[a]	10 ppm	0.1 ppm

[a] See description of modified devices in Section V,F.

Such films have been observed to be sensitive to a number of other gases, for example, carbon monoxide and ethanol. The mechanism in this case is probably penetration of complete gas molecules through pores in the metal to the metal–silicon dioxide interface, where they form dipoles. Another way to change the palladium gate is to combine it with other catalysts. We recently made reproducible ammonia sensors in this way by evaporating a thin layer of other catalysts onto the structure (Spetz et al., 1983; Winquist et al., 1983; Spetz et al., 1984).

D. PRACTICAL DEVICES

Gas sensors based on Pd-gate MOS devices can be made in different ways. The simplest device is the MOS capacitor; a more complex device is an integrated circuit containing one Pd-gate MOS transistor, one diffused resistor, and one pn diode, all in one piece of silicon. The resistor is used as a heater and the diode as a temperature sensor in this circuit.

An MOS capacitor device is easy to make (Lundström et al., 1977). To fabricate a device in p-type silicon, the silicon wafer is first oxidized in dry oxygen at 1200°C to a thickness of 100 nm. The oxide is etched away from the back of the wafer and aluminum is evaporated there. The aluminum layer may then be sintered at 500°C in forming gas. Then palladium is evaporated onto the front side of the wafer through a mechanical mask, producing Pd dots about 1 mm² in area. The Pd layer is typically 50–100 nm thick. In order to avoid radiation damage, thermal evaporation is preferred to electron beam evaporation. Palladium normally has very weak adhesion to silicon dioxide, which causes the Pd film to peel off during fabrication (Shivaraman and Svensson, 1976) and form blisters during use (Armgarth and Nylander, 1982). Heat treatment in air at 200°C (Shivaraman and Svensson, 1976) and the use of thicker Pd films (> 400 nm) (Armgarth and Nylander, 1982) have been shown to solve these problems. To make a complete sensor out of the capacitor, it is combined with a heater and a temperature sensor to keep the device at an appropriate, constant temperature. This may be arranged by mounting the capacitor on a thick-film substrate with a printed heater resistor and a mounted temperature sensor. When using the sensor, a temperature-regulating circuit is used to keep the temperature constant and a capacitance meter is used to measure the capacitance. Here one may either use the capacitance as a measure of the gas concentration (Steele *et al.*, 1976) or use a regulator to keep the capacitance constant by varying the bias voltage. In the latter case the bias voltage is used as a measure of the gas concentration (Lundström *et al.*, 1977).

The first devices made were simple *n*-channel MOS transistors (Lundström *et al.*, 1975a,b). They had very thin oxide and palladium layers, about 10 nm each, and were made by a special fabrication process. Complete sensors were made by mounting these transistors on transistor headers with an externally attached heater and temperature sensor (Stiblert and Svensson, 1975). When using a transistor, the change in threshold voltage is easily detected directly, as shown in Fig. 9.

A more complex device is the integrated circuit mentioned above. The fabrication sequence follows that of standard integrated circuits except for the last step, the palladium layer. Figure 14 shows a layout of such a circuit and Fig. 15 shows its schematic and use in an instrument. In the development of practical devices several problems may arise. Apart from the normal problems in semiconductor processing, the Pd-gate MOS sensor involves some special considerations. The problem of Pd adherence to the oxide has been discussed above. Here new ideas had to be used, as the normal method for increasing adherence—for example, with an intermediate chromium layer—changes the gas sensitivity of the device.

In order to minimize drift in the finished devices, all normal MOS drift phenomena must be minimized. This means mainly that the silicon dioxide must be of the highest quality, free of alkali ions and radiation damage. A very thin oxide also minimizes drift phenomena, as many such phenomena are proportional to the oxide thickness (see also Section VI).

Finally, the properties of the metal gate layer are crucial for these devices. The Pd layer must be fabricated with great care to ensure reproducible and stable results. Metal problems are discussed further in Section V.

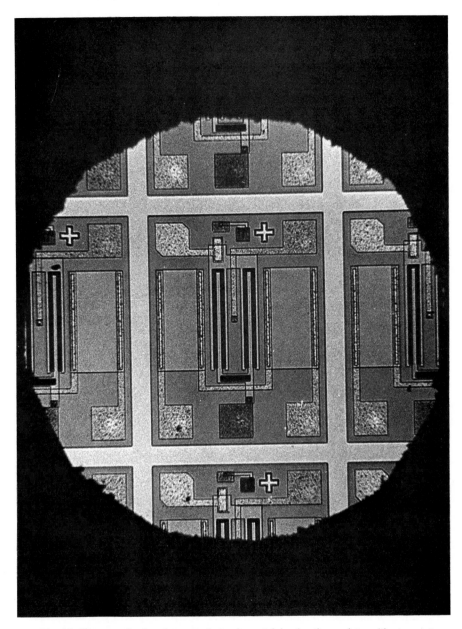

Fig. 14. Photograph of an integrated circuit containing heating resistors (the two outer rectangles), a diode for temperature control (in the middle), and a Pd-gate MOS transistor (in two parts around the middle). The transistor gate is shorted to its drain (upper middle contact). The chip size is about 0.5 × 0.7 mm.

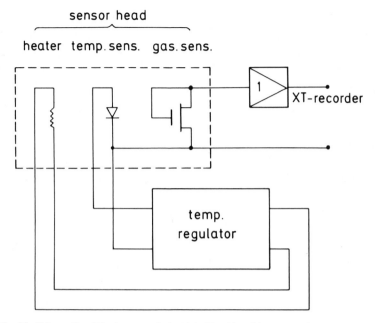

Fig. 15. Schematic of the integrated circuit in Fig. 14 and its connection to the measuring equipment.

E. Sensor Performance

The practical performance of the MOS sensors is controlled mainly by three crucial properties: sensitivity, selectivity, and stability. These properties, although correlated, will be discussed one at a time.

The practical sensitivity of the sensor is given by the practical detectable voltage change. This is limited mainly by disturbances of different kinds, such as noise and drift. In the case of Pd-gate devices temperature drift and device drift dominate. The temperature drift of most semiconductor devices is of the order of one or a few millivolts per degree (reflecting the rate of change of the Fermi level in the semiconductor). A realistic temperature stability may be ±1° or somewhat better, giving rise to a temperature drift of one or a few millivolts. By device drift we mean slow changes in the voltage at zero gas concentration, depending on aging and so on. This drift depends strongly on time. For short times (minutes) the device drift may be considered less than 1 mV.

Keeping the above discussion in mind, we may consider the different sensitivities shown in Table I for a 10-mV detection limit as conservative estimates. The sensitivity to hydrogen must be considered good compared to that of other sensors. The detection limit of 0.5 ppm of gas in air is about the same as that of a helium leak detector used in the same manner. It is also about equal to the natural concentration of hydrogen in air. In an inert atmosphere the sensitivity is very high, 0.03 ppb or about 10^{-8} torr—comparable to many high-vacuum instruments. Recent experiments have shown that the device is useful down to 10^{-10} torr H_2 in ultrahigh vacuum (Petersson et al., 1982a,b). It is also possible to increase the resolution of the device by using it in a pulsed gas mixing system in which the device is exposed periodically to a reference gas and the gas to be tested, respectively. The observed sensitivities to hydrogen sulfide and ammonia are somewhat less than the sensitivity to hydrogen, but still reasonably good. The values can, for example, be compared to the safety limits, which are 20 ppm for hydrogen sulfide and 50 ppm for ammonia (8 h exposure). The sensitivity is thus good enough for controlling health hazards with these gases.

The selectivity of the Pd-gate MOS sensor is also quite good. Most other solid state gas sensors are sensitive to broad classes of gases, for example, to all combustible gases. The Pd-gate device is, however, sensitive to only a few hydrogen compounds. Although its selectivity has not been fully investigated, we know that it is sensitive to hydrogen, hydrogen sulfide, and, in some cases, ammonia. We also know that it is not sensitive to simple hydrocarbons, simple alcohols, carbon monoxide, or water vapor.

The situation may be more complicated in the presence of more than one gas. The sensor may, for example, react to a mixture of carbon monoxide and water vapor because of the reaction

$$CO + H_2O \rightarrow H_2 + CO_2 \tag{15}$$

It may also respond to some gases in the presence of a background hydrogen pressure, for example, with decreased hydrogen sensitivity as a result of adsorption of the other gas.

Three main sources of instability in these devices are zero-point drift (slow change of the voltage corresponding to zero gas concentrations), hydrogen-induced drift (an extra slow hydrogen response), and inactivation in oxygen (slowness of the device after storage in oxygen).

The zero-point drift (during storage in, e.g., air) is due to normal drift phenomena in MOS structures and can be minimized through suitable processing. It is thus favorable to work with n-channel devices—that is, p-substrates—to avoid the so-called negative bias instability (Jeppson and Svensson, 1977). Furthermore, the contribution to the threshold voltage from the drift, either ionic or electronic, is proportional to the oxide thickness. It is therefore possible to reduce the influence of these phenomena through a decrease in oxide thickness, because the contribution

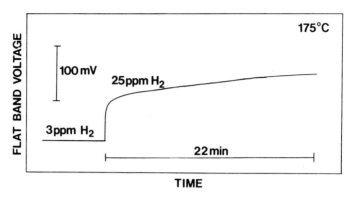

Fig. 16. Voltage shift versus time for a step in hydrogen concentration.

from the hydrogen dipoles is independent of oxide thickness. We do not consider the normal MOS drift to be a large problem in Pd-MOS devices.

Hydrogen-induced drift (Svensson, 1980; Lundström and Söderberg, 1981–1982) is demonstrated in Fig. 16 at relatively long times after an increase in hydrogen concentration. Initially there is a normal fast voltage shift ΔV_N. Later, however, the voltage continues to change slowly for quite long times (in Fig. 16 the process is somewhat speeded up by a higher than normal temperature). The slow voltage shift may be 0.5 V. When hydrogen is removed again the normal shift ΔV_N will disappear within 10 s but the slow shift will remain for hours; it seems as though the voltage shift will not return to its original value. That is why we also term this phenomenon hysteresis. The effect is obviously very disturbing when the device is used as a gas sensor. As in the case of device drift, it can be overcome by frequent recalibration or by the use of a pulsed gas stream. The slow shift and the fast normal shift are additive. We have observed that the slow shift disappears if the device is stored in an excess of hydrogen. Figure 17 shows the result of a long-term experiment in which the device was stored in oxygen or hydrogen and subjected to short pulses of a hydrogen–oxygen mixture at intervals (Söderberg, 1983). During oxygen storage a slow discharge takes place, but the reverse, a slow charging, is not observed during storage in hydrogen. This is because in hydrogen the process responsible for the slow effect is rapidly saturated.

The origin of the hydrogen-induced drift is found at the metal–silicon dioxide interface. It has been shown that it is caused by slow hydrogen adsorption sites in the oxide (Nylander *et al.*, 1981; Nylander *et al.*, 1984).

Another practical observation is that if the device is stored in hydrogen (or in argon) and used as an oxygen sensor it will be extremely stable between the oxygen exposures (Lundström and Söderberg, 1981; Söder-

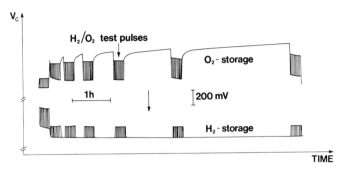

Fig. 17. Long-term experiments indicating the stability of a Pd-MOS structure in different environments, hydrogen and oxygen. At intervals the structure was subjected to (11) pulses of a hydrogen–oxygen mixture (260 Pa–15,000 Pa) with argon as the carrier gas. The two curves have been displaced for clarity. The arrow indicates the direction of the hydrogen-induced voltage shift. Device temperature was 150°C. (Data from Söderberg, 1983.)

berg, 1983). Experimental results supporting this are shown in Fig. 17. Recent experiments indicate that the hydrogen-induced drift can be minimized or avoided: by introducing a thin layer of aluminum oxide between the palladium and silicon dioxide remarkably stable devices were made (Armgarth and Nylander, 1981; Hua *et al.*, 1984a). Virtually no drift was observed in such a device after 15 min in 100 ppm hydrogen in oxygen, compared to about 0.2 V of drift in a normal device (Fig. 18). The results in Fig. 18 indicate that it should be possible to find a stable Pd–insulator interface. Investigations of different Pd–insulator interfaces have recently been extended to $Pd-Si_3N_4$ and $Pd-Ta_2O_5$, which also show better stability than the $Pd-SiO_2$ interface (Dobos *et al.*, 1984).

Oxygen deactivation is demonstrated in Fig. 19 (Lundström and Söderberg, 1981–1982). If a Pd-MOS device is kept in air or in oxygen without hydrogen for a long time it responds very slowly the first time it is exposed to hydrogen again. After a few hydrogen doses, however, it recovers its normal response. The phenomenon probably indicates oxygen adsorption or palladium oxidation in oxygen and subsequent reduction in the presence of hydrogen. It suggests that the device should be very difficult to use without calibration before each measurement. However, it has been observed that if the device is stored hot in normal air the background hydrogen in the air keeps the surface clean enough to yield a reasonable response time. It should also be noted that only the response time and not the steady-state value is affected by the storage.

An important practical parameter is the dynamic range of the sensor, that is, the range of gas concentrations that can be measured. Assuming Eq. (13) to be valid, we may calculate the relative differential sensitivity

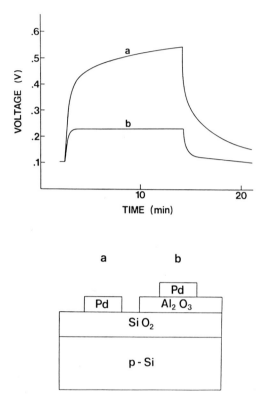

Fig. 18. Illustration of the properties of Pd–Al₂O₃–SiO₂–Si structures. The upper drawing shows the flat-band voltage shift when the test structures (shown below) are subjected to a change in hydrogen pressure from 10 to 100 ppm in oxygen and back again after about 12 min. Device temperature was 150°C. (Data from Armgarth and Nylander, 1981.)

$$p \, \frac{\delta \Delta V}{\delta p} = \frac{\Delta V_{\text{max}} \, C \, \sqrt{p}}{2(1 + C \sqrt{p})^2} \tag{16}$$

Let us assume that we want to detect at least a 50% change in gas concentration. With a detection limit of 10 mV change ($\delta \Delta V$), a maximum shift of 0.5 V (ΔV_{max}), and $\delta p/p = 0.5$, we find a dynamic range of about $0.1 < C\sqrt{p} < 10$, corresponding to a relative pressure range of 1 : 10,000. This is quite a large range, chiefly as a result of the square root in Eq. (13). The practical range may be even larger because new adsorption mechanisms may set in when the normal one starts to saturate (Lundström and Söderberg, 1981–1982). Furthermore, a 10-mV detection limit is a conservative estimate.

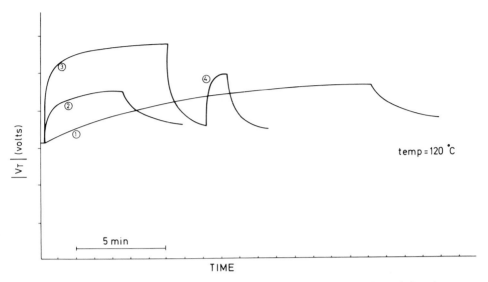

Fig. 19. Response to hydrogen in air when a device has been kept in clean air for a long time. (1) First exposure, 40 ppm H_2; (2) second exposure, 40 ppm H_2; (3) third exposure, 210 ppm H_2; (4) fourth exposure, 40 ppm H_2. One division on the V_T axis is 100 mV.

Although there are still some problems with the Pd-MOS devices, we regard them as useful in many applications; we shall return to this discussion in Section VII.

IV. Schottky Barrier Devices

A. The Basic Metal–Semiconductor Diode

Metal-semiconductor diodes can be formed in several ways. A metal film can be evaporated directly onto a clean or oxidized semiconductor surface, or it can be evaporated and thereafter reacted with the semiconductor by heat treatment. A diode formed in such a way is called a Schottky diode (this name is also used if there is a very thin oxide layer between the metal and the semiconductor; see Fig. 20).

In a Schottky diode a barrier is normally formed between the metal and the semiconductor. If this barrier is large enough the diode will be rectifying, that is, it will exhibit an asymmetric current–voltage curve, as shown in Fig. 20. For the simplest case the barrier formed between the metal and an n-type semiconductor will be given by (Sze, 1981)

$$\psi = \phi_M - \chi_s \qquad (17)$$

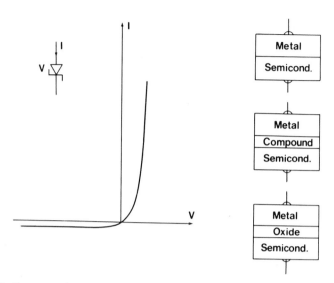

Fig. 20. Current–voltage characteristics of a Schottky barrier diode and three types of such diodes. The three types could be, for example, palladium–silicon, palladium–palladium silicide–silicon, and palladium–silicon dioxide–silicon structures.

(See also the band diagram in Fig. 21a.) The barrier height is proportional to the metal work function, which indicates that this device also can measure gases by monitoring the metal work function. Equation (17) is, however, valid only for strongly ionic semiconductors, such as oxides, and not, for example, for silicon (Kurtin *et al.*, 1969). This is because large amounts of surface states are formed at the metal–semiconductor interface in the case of covalent materials. The barrier height will then be controlled by these states instead of the metal work function. Most silicon Schottky diodes thus have a fixed barrier height of about two-thirds of the band gap for *n*-type material. In the case of oxidized silicon, however, a thin oxide layer will remove most of these surface states but can still be thin enough to permit electron tunneling. For an "MOS" Schottky diode the barrier height will be given approximately by Eq. (17). The Schottky diode capacitance depends on voltage according to (Sze, 1981)

$$C = \frac{k_1}{\sqrt{(\psi - \phi_F - V)}} \tag{18}$$

where k_1 is a constant, ψ is the metal–semiconductor barrier height, and V is the applied voltage ($V > 0$ for an *n*-type semiconductor when the metal is positive); ϕ_F is defined in Fig. 3. The capacitance–voltage characteristic

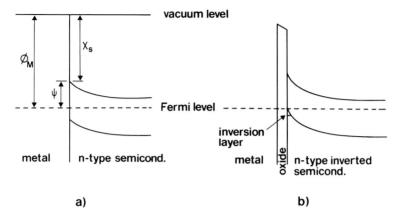

Fig. 21. Band diagrams of (a) a metal-semiconductor diode and (b) a metal–oxide–semiconductor diode with an inversion layer.

moves along the voltage axis if the metal work function (and therefore ψ) changes, just as in the case of the MOS capacitor.

The diode current is given by (Sze, 1981)

$$I = k_2 \, e^{-q\psi/kT}(e^{qV/kT} - 1) \qquad (19)$$

where k_2 is a constant and kT/q is the thermal voltage. For the backbiased diode ($V \ll 0$) the current can be written

$$I = -k_2 \, e^{-q\psi/kT} \qquad (20)$$

For a practical device the voltage dependence is often somewhat different from that given by Eq. (19) for the ideal case. To account for this a quality factor n is often introduced in the expression for the current

$$I = k_2 \, e^{-q\psi/kT}(e^{qV/nkT} - 1) \qquad (21)$$

where n is of the order of one. Again, the situation is similar to that of the MOS device, that is, the current–voltage characteristic moves along the voltage axis upon a change in barrier height or metal work function.

For an MOS Schottky barrier the above equations are normally approximately valid. There is, however, one important exception. When an inversion layer is formed in the semiconductor surface beneath the oxide, as in Fig. 21b (compare the inversion layer used as a conducting channel in the MOS transistor), the n-type semiconductor is converted to p-type at the surface. The device thus looks more like a pn diode in series with a non-current-limiting tunnel junction. The diode characteristic is independent of the metal work function as long as the inversion condition persists (Green *et al.*, 1974).

B. Gas-Sensitive Metal–Semiconductor Schottky Diodes

A simple palladium–silicon Schottky barrier diode is not sensitive to hydrogen gas according to the above description. This has been confirmed by experiments (Yamamoto *et al.*, 1980, 1981; Ruths *et al.*, 1981). It has also been observed that palladium silicide–silicon diodes are insensitive to hydrogen (Fonash *et al.*, 1982).

The first successful gas-sensitive Schottky barrier diode was reported by Steele and MacIver (1976). They made a Pd–CdS diode that showed hydrogen sensitivity in the concentration range 500–5000 ppm hydrogen in air at 25–50°C. The response time reported was quite long, 10–15 min. Since then several investigators have described gas-sensitive metal–semi-conductor diodes (Ito, 1979; Yamamoto *et al.*, 1980, 1981; Harris, 1980; Yousuf *et al.*, 1982; Poteat *et al.*, 1983).

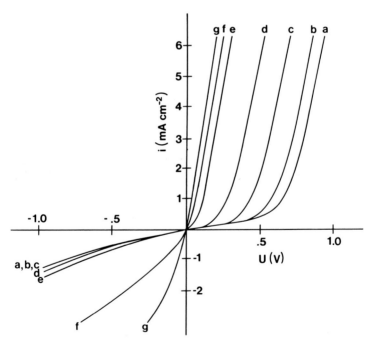

Fig. 22. Current–voltage characteristic of a Pd–TiO₂ Schottky diode in air with different hydrogen concentrations. The partial pressures are (a) 0, (b) 14 ppm, (c) 140 ppm, (d) 1400 ppm, (e) 7150 ppm, (f) 1%, and (g) 1.5%. The temperature was 25°C. (From Yamamoto *et al.*, 1980. © North-Holland Physics Publishing, Amsterdam, 1980.)

With Pd–ZnO Schottky barriers, Ito (1979) observed barrier height changes by all the methods mentioned above—capacitance, reverse current, and forward current measurements. The device was a good hydrogen sensor down to room temperature in a concentration range of 0–600 ppm hydrogen in air. Negative barrier height changes of up to 0.3 eV were observed; the response time was of the order of 1 min.

Pd–TiO$_2$ Schottky barrier diodes were investigated by Yamamoto *et al.* (1980) and Pt–TiO$_2$ structures by Harris (1980). Figure 22 shows experimental current–voltage curves for Pd–TiO$_2$ diodes at different hydrogen concentrations; in Fig. 23 the estimated change in barrier height is plotted against hydrogen concentration. It is clear from the current–voltage characteristic how the barrier height starts at a reasonably high value to give a nice rectifying characteristic and then decreases enough to make the diode nearly ohmic. The initial barrier height was 0.67 eV, the same value as expected from a direct measurement of the work function difference between the two materials.

The device shows very promising behavior as a hydrogen sensor. At 80°C the response time is about 1 min and the sensitivity range may be from about 0.1 ppm to about 2%. It is almost insensitive to humidity and has no sensitivity to other gases (such as hydrocarbons) at temperatures below 60°C. At higher temperatures, however, it becomes sensitive to other reducing gases such as CO, C$_3$H$_6$, or C$_2$H$_5$OH. The Pd films on the TiO$_2$ structures were probably porous, which may explain the sensitivity

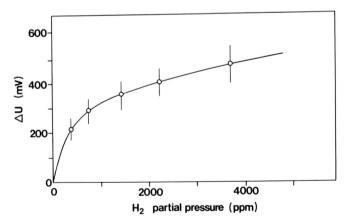

Fig. 23. Estimated change in barrier height of a Pd–TiO$_2$ Schottky diode at 25°C. (From Yamamoto *et al.*, 1980. © North-Holland Physics Publishing, Amsterdam, 1980.)

to gases other than H_2 (see Section III, C). Yamamoto *et al.* (1980) investigated a number of other metal–semiconductor combinations: Pt, Au, Ni, Al, Cu, Mg, and Zn on TiO_2 and Pd on ZnO, GaP, CdS, and Si. These devices either could not be used as sensors or were inferior to the Pd–TiO_2 device.

Yousuf *et al.* (1982) and Poteat *et al.* (1983) have made extensive investigations of Schottky barrier devices. They conclude that in these devices the surface state density of the semiconductor is also influenced by hydrogen, which was first suggested by Ruths *et al.* (1981). The change in surface state density is observed as a change in the quality factor n in Eq. (21).

C. Gas-Sensitive MOS Schottky Diodes

MOS Schottky barrier diodes made with palladium metal on oxidized n-type silicon have been shown to be gas sensitive in a manner similar to the normal MOS devices (Shivaraman *et al.*, 1976; Keramati and Zemel, 1978, 1982a,b; Ruths *et al.*, 1981; Ito, 1981; Fonash *et al.*, 1982). The behavior is, however, disturbed by the inversion layer formation mentioned above (Shivaraman *et al.*, 1976).

One way to measure the barrier height is to measure the reverse current and take its logarithm, from Eq. (20)

$$\frac{kT}{q} \ln|I| = \text{const} - \psi \tag{22}$$

Figure 24 shows $(kT/q) \ln|I|$ versus the hydrogen partial pressure, taken from measurements on a Pd–oxidized Si diode with 1.5-nm-thick oxide (Shivaraman *et al.*, 1976). This result indicates that an inversion layer exists at low hydrogen pressures, making the device insensitive to changes in barrier height (compare with the dashed line in Fig. 24, which is the expected barrier height change from measurements on conventional MOS devices).

At higher pressures the inversion layer disappears and the device becomes hydrogen-sensitive for partial pressures above about 5 Pa, corresponding to a barrier height change of about 0.2 eV. A theoretical estimate of the barrier height between hydrogen-free Pd and silicon indicates a value of about 1 eV [Eq. (17) with $\phi_M = 5.1$ eV and $\chi_s = 4.1$ eV]. This leaves only 0.1 eV between the metal Fermi level and the silicon valence band, strongly suggesting inversion layer formation. After ϕ_M is lowered by 0.2 eV the Fermi level will be 0.3 eV above the silicon valence band, indicating disappearance of the inversion layer.

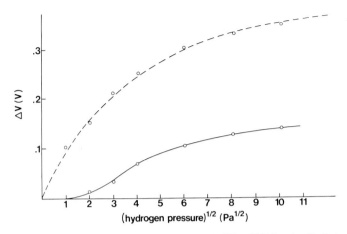

Fig. 24. Estimated change in barrier height of a Pd–SiO$_2$–Si Schottky diode (solid line) compared to expected change from Pd-gate MOS transistor measurements (dashed line). The temperature was 150°C. (From Shivaraman *et al.*, 1976.)

In addition to the expected hydrogen-sensitive barrier height, Keramati and Zemel (1978, 1982a,b) made observations which they attributed to hydrogen induced changes in the density of silicon dioxide–silicon interface traps. On the other hand, it has been suggested that hydrogen does not induce changes in the interface trap density in MOS capacitors (Poteat and Lalevic, 1982; Poteat *et al.*, 1983; Fonash *et al.*, 1982). The thickness of the oxide is probably an important factor in this respect. Hydrogen induced changes in the interface trap density have been observed in devices with thin oxides (Fare and Zemel, 1984).

D. SWITCHING DEVICES

New types of device structures with palladium gates have also been proposed. One interesting structure is the Pd-gated switching device developed by Kawamura and Yamamoto (1983a,b). This device, which is a Pd–thin SiO$_2$–*n*-*p*-Si structure, has a turn-on voltage that depends on the ambient hydrogen concentration. Through a suitable choice of bias voltage the device can be set to switch from a low- to a high-conductance state at a given hydrogen concentration.

V. Properties of the Catalytic Metal

A. THE THREE ROLES OF THE METAL

The metal has three roles in metal gate gas sensors: it dissociates the incoming gas through its catalytic action, it transports hydrogen atoms to the metal–oxide interface, and it adsorbs the hydrogen atoms at this interface as detectable dipoles.

The first role means that hydrogen molecules, hydrogen sulfide molecules, and so on are dissociated so that hydrogen atoms are formed and dissolved in the metal. This catalytic action is characteristic of metals like palladium and platinum. Other metals or metal oxides (e.g., palladium oxide) can also be active catalysts in this way. The outer surface is also catalytic for other surface reactions of importance for the sensor; for instance, it can catalyze oxidation of hydrogen by oxygen or chlorine, or catalyze oxidation of sulfur by oxygen.

The second role of the metal is to transport the hydrogen atoms through the metal layer. Hydrogen is unique in this respect, because the hydrogen ion is a bare proton and thus much smaller than any other ion or atom. It is reasonable to believe that no other atom will diffuse as rapidly as hydrogen through a rigid metal film. The diffusion of hydrogen atoms is very fast in palladium as well as in most metals. The large solubility of hydrogen in palladium (Lewis, 1967), however, makes palladium unique.

Of course, if a porous metal layer is used as the gate material any gas can diffuse directly through the metal layer (without catalytic dissociation).

The third role of the metal layer is to adsorb the diffused atoms (or molecules) at the metal–support (insulator, semiconductor) interface as dipoles. The dipole layer formed there will be detectable by the semiconductor as a change in the semiconductor surface field. By using a surface field-sensitive device, such as an MOS capacitor, an MOS transistor, or a Schottky diode, a measurable signal may be generated from this change in field.

The amount of hydrogen adsorbed onto the Pd–insulator or Pd–semiconductor interface depends not only on the hydrogen pressure but also on the chemical reactions on the metal surface; these devices become extremely sensitive to hydrogen in an inert atmosphere (see Table I). Possible surface reactions are discussed below, starting with the simplest possible case, namely with hydrogen in an inert atmosphere (argon).

B. SURFACE REACTIONS IN AN INERT ATMOSPHERE

In an inert atmosphere there are two important steps (illustrated in Fig. 25), namely dissociation and association of hydrogen on the surface

$$H_2 \underset{d_1}{\overset{c_1}{\rightleftharpoons}} 2 H_a \qquad (23)$$

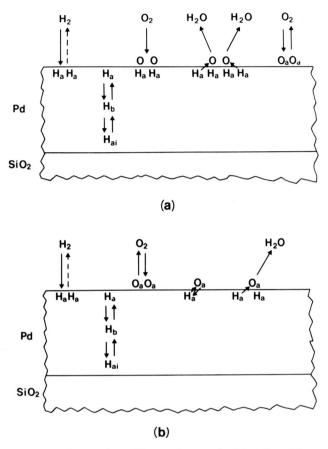

Fig. 25. Schematic picture of possible reactions on the Pd surface. The reactions are discussed in the text.

and transfer of hydrogen to the interface and back via bulk hydrogen

$$H_a \underset{d_e}{\overset{c_e}{\rightleftharpoons}} H_b \underset{c_i}{\overset{d_i}{\rightleftharpoons}} H_{ai} \tag{24}$$

where the c's and d's are rate constants.

If we introduce the numbers of adsorption sites N_e and N_i and the numbers of adsorbed hydrogen atoms n_e and n_i at the surface and interface, respectively, we find that in the steady state

$$\frac{n_i}{N_i - n_i} = \frac{c_e d_i}{d_e c_i} \frac{n_e}{N_e - n_e} = \frac{c_e d_i}{d_e c_i} \sqrt{\frac{c_1 P_{H_2}}{d_1}} \equiv k_0 \sqrt{P_{H_2}} \tag{25}$$

This equation expresses a true equilibrium isotherm, the well-known Langmuir isotherm (Bond, 1974). The connection between n_i and the voltage shift ΔV is obtained by assuming that the shift is proportional to the coverage of hydrogen atoms at the interface

$$\Delta V = \Delta V_{max}\theta \tag{26}$$

where $\theta \equiv n_i/N_i$ and ΔV_{max} is the maximum obtainable voltage shift. Some experimental results showing the voltage shift for hydrogen in argon are given in Fig. 26. The high temperatures are necessary because the Pd-MOS structures are highly sensitive to hydrogen at lower temperatures, a consequence of the large exothermic heat of adsorption of hydrogen on palladium. From the temperature dependence of k_0 calculated from the experimental results with Eqs. (25) and (26) we obtain the value 1.1–1.2 eV per H_2 molecule (Lundström et al., 1977), which is close to the value found for surface adsorption of hydrogen, 1.1 eV per H_2 molecule (Bond, 1974).

C. Surface Reactions in Oxygen

The situation on the metal surface is very complex even in a simple mixture of hydrogen and oxygen. The reactions that take place on the Pd surface are not known in detail. In an excess of oxygen the surface is probably covered by an oxide layer or a layer of chemisorbed oxygen atoms. Hydrogen adsorption takes place in or through such a layer. Furthermore, water molecules can be produced by at least two different routes, as illustrated in Fig. 25. The behavior of hydrogen is, in principle, described by equations similar to Eqs. (23) and (24). To these we must add equations describing the behavior of oxygen and the reactions between hydrogen and oxygen.

In one possible reaction sequence (Fig. 25a), water production takes place via oxygen molecules that are dissociated on adsorbed hydrogen atoms

$$O_2 + 2\ H_a \xrightarrow{\ c_2\ } 2\ (OH)_a \tag{27}$$

and

$$2\ H_a + 2\ (OH)_a \xrightarrow{\ c_3\ } 2\ H_2O\ (gas) \tag{28}$$

In Eqs. (27) and (28) it is not clear whether $2\ (OH)_a$ exists as two adsorbed OH groups or as H_2O_2. In the steady state

$$\frac{n_i}{N_i - n_i} = \frac{c_e d_i}{d_e c_i}\ \frac{n_e}{N_e - n_e} = \frac{c_e d_i}{d_e c_i}\ \sqrt{\frac{c_1 P_{H_2}}{2c_2 P_{O_2}}} \tag{29}$$

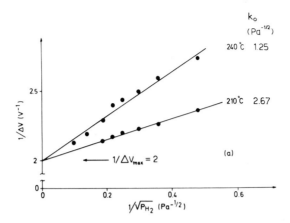

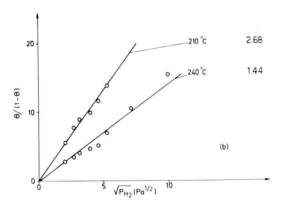

Fig. 26. Experimental results showing relation of voltage shift and hydrogen coverage to hydrogen partial pressure for hydrogen in argon at two different temperatures. The data are plotted in two ways to demonstrate the agreement with Eqs. (25) and (26). (a) $1/\Delta V$ is expected to be of the form

$$\frac{1}{\Delta V} - \frac{1}{\Delta V_{max}} = \frac{1}{\Delta V_{max}} \frac{1}{k_0} \sqrt{\frac{1}{P_{H_2}}}$$

Extrapolation to $1/\sqrt{P_{H_2}} = 0$ gives

$$1/\Delta V_{max} = 2 \quad \text{or} \quad \Delta V_{max} = 0.5 \text{ V}$$

(b) From Eq. (25)

$$\frac{\theta}{1 - \theta} = k_0 \sqrt{P_{H_2}}$$

$\theta \equiv n_i/N_i$ is the coverage of hydrogen atoms at the interface.

if the fundamental back reaction $2H_a \rightarrow H_2$ is assumed to be slow compared to water production, an assumption that is justified by experimental results.

In an alternative reaction scheme (Fig. 25b) water is produced via adsorbed oxygen

$$O_2 \underset{d_4}{\overset{c_4}{\rightleftarrows}} 2 O_a \qquad (30)$$

$$H_a + O_a \underset{d_5}{\overset{c_5}{\rightleftarrows}} (OH)_a \qquad (31)$$

$$(OH)_a + H_a \overset{c_6}{\longrightarrow} H_2O \text{ (gas)} \qquad (32)$$

In this case there are several possible models, depending on the assumptions about the adsorption sites and the relative sizes of the rate constants. In one special case, leading to a simple relation, it is assumed that c_6 is small compared to c_5, d_5, and d_4; that is, the water production is slow. Furthermore, it is assumed that oxygen is adsorbed on special sites (N_o) that do not compete with hydrogen adsorption (other than through simple geometric blocking; see below). With these assumptions we find

$$\frac{n_o}{N_o - n_o} \approx \sqrt{\frac{c_4 P_{O_2}}{d_4}} \qquad (33)$$

and

$$\frac{n_i}{N_i - n_i} = \frac{c_e d_i}{d_e c_i} \frac{n_e}{N_e - n_{OH} - n_e} = \frac{c_e d_i}{d_e c_i} \sqrt{\frac{c_1 P_{H_2}}{2 c_6 c_5 n_o / d_5}}$$

$$= \frac{c_e d_i}{d_e c_i} \sqrt{\frac{d_5 c_1 P_{H_2}(1 + \sqrt{c_4 P_{O_2}/d_4})}{2 c_6 c_5 N_o \sqrt{c_4 P_{O_2}/d_4}}} \qquad (34)$$

which is quite different from Eq. (29). A further discussion of these two models is found in Lundström and Söderberg (1981–1982).

Experimentally, oxygen dependences of the form $P_{H_2}^{1/2}/P_{O_2}^{\alpha}$ with $\alpha = \frac{1}{4}-\frac{1}{2}$ are found. The empirical relation, Eq. (14), is therefore not generally valid. A careful comparison between theory and experiments has been hampered by the slow hydrogen adsorption sites discussed in Section III,E. The experimentally observed oxygen dependence is partly determined by these slow sites. We believe, however, that the fast and stable hydrogen response of, for example, Pd–Al$_2$O$_3$ structures can be used to shed light on the oxygen dependence of Pd-gate hydrogen sensors. It is interesting to note that a change in the oxygen dependence with temperature from $P_{O_2}^{-1/4}$ to $P_{O_2}^{-1/2}$ has been observed on Pd-MAOS structures (Hua et al., 1984b).

Another experimental observation related to the adsorption of oxygen is that when hydrogen is introduced the initial rate of change of ΔV is dependent on the oxygen pressure. This can be understood if oxygen atoms occupy the same type of adsorption sites as hydrogen, or if adsorbed oxygen blocks hydrogen adsorption by decreasing the available area (or number of sites) for hydrogen adsorption. In both cases we expect the initial rate of change of ΔV to be given by

$$\frac{d(\Delta V)}{dt} \approx \text{const } P_{H_2}(N - n_0) \tag{35}$$

where

$$N - n_0 = \frac{N}{1 + \sqrt{c_4 P_{O_2}/d_4}} \tag{36}$$

and $N = N_e$ or N_0 for the two different cases discussed above.

Experimental results obtained by Plihal (1977) and by Lundström and Söderberg (1981–1982) follow Eqs. (35) and (36) closely. An example is shown in Fig. 27, where we have plotted the experimentally observed rate

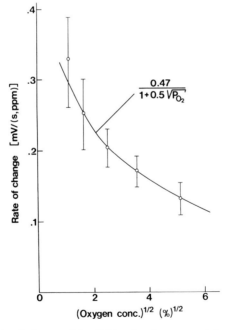

Fig. 27. Initial rate of change of the flat-band voltage upon introduction of hydrogen versus square root of oxygen concentration. Hydrogen was introduced in a gas stream containing oxygen (in argon). The measured rate was normalized with the hydrogen concentration (around 400 ppm) since it was not possible to keep the concentration exactly constant during the whole experiment. The data were fitted to Eqs. (35) and (36). The temperature of the sample was 152°C.

of change against partial oxygen pressure (or concentration) and compared it with the theoretical expression. We also performed some experiments in which the temperature dependence of the initial rate of change was determined (Lundström *et al.*, 1977), and found that the dissociation of hydrogen exhibits a temperature dependence of the form $\exp(-E_1/kT)$ with $E_1 \approx 0.25$–0.3 eV.

The complexity of the surface reactions is demonstrated by the experimental results shown in Fig. 28. In this case the Pd-MOS structure was kept in a small amount of hydrogen in argon and subjected to oxygen pulses 2 min long. In Fig. 28 ΔV is the change in V_{FB} during the oxygen pulse. Hysteresis is observed at the higher hydrogen pressures and is connected with a turn on phenomenon, which also shows up in transient records of the response to oxygen (Lundström and Söderberg, 1981–1982). The hysteresis is due to a change in the number of reaction sites on the surface of the Pd film when the structure is subjected to oxygen.

D. FUNDAMENTAL STUDIES

To learn more about the catalytic metal at atmospheric pressures, we measured the surface potential of the Pd film. This was done with a vibrating metal electrode above the Pd surface, a so-called Kelvin probe (Söderberg *et al.*, 1980; Söderberg and Lundström, 1980, 1983; Lundström and Söderberg, 1981–1982, 1982; Söderberg, 1983). Simultaneous measurements of the surface potential ϕ_e and the interface potential ϕ_i

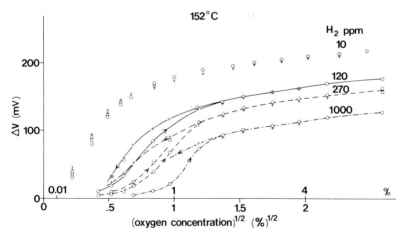

Fig. 28. Change in flat-band voltage during a 2-min oxygen pulse at the concentration shown on the abscissa, with hydrogen concentration as a parameter. The arrows indicate increasing and decreasing oxygen concentrations, respectively.

revealed a number of interesting facts (note that $\Delta\phi_i \equiv \Delta V$); of particular note:

1. For normal Pd-MOS structures there is a "quasi-equilibrium" between adsorbed hydrogen atoms on the surface and at the interface during transient conditions.
2. Oxygen creates a large surface potential change with a direction opposite to that created by hydrogen.
3. There are several types of hydrogen adsorption sites on the surface and at the interface.
4. There are adsorption sites at the palladium–silicon dioxide interface or in the oxide that have no counterparts on the surface.
5. The inactivation of the response observed after storage in oxygen for a long time is related to the surface.
6. For a porous metal film the surface and interface potential changes are about the same in almost any gas mixture.

An interesting observation has been made regarding the interaction between hydrogen and oxygen on the palladium surface: if the surface is "cleaned" through exposure to hydrogen, kept in argon, and subjected to short pulses of a hydrogen–oxygen mixture, a transition from a hydrogen- to an oxygen-covered surface can be observed below a certain P_{H_2}/P_{O_2} ratio (Söderberg and Lundström, 1983; Söderberg, 1983). Furthermore, a new type of hydrogen adsorption site appears on the oxygen-covered surface. The transition from a hydrogen- to an oxygen-covered surface is explained by competition between hydrogen and oxygen atoms (and OH groups) for the same type of adsorption sites on the surface. In normal operation in air the surface is therefore certainly covered with oxygen. The exact connection between the results discussed above and the sensitivity to hydrogen in air is, however, not known at present.

Palladium surfaces that have been exposed to air are in many respects different from the clean metal surfaces studied by spectroscopic methods in ultrahigh vacuum (UHV) systems. For example, a contaminated Pd surface has a very low sticking coefficient for hydrogen; that is, the probability that a hydrogen molecule will be adsorbed (and dissociated) when it hits the contaminated metal surface is very low. Nevertheless, careful studies in UHV systems and the use of surface analytical tools to determine the properties of the surface and to probe surface reactions are probably fruitful ways to learn more about the catalytic part of the sensor. We have therefore started some studies in a UHV system (Petersson *et al.*, 1982a,b, 1984a,b; Dannetun *et al.*, 1984).

There appears to be a large difference in UHV between the surface and the interface at operating temperatures of 100–150°C. No distinct hydrogen adsorption sites are visible at a clean surface, although they are observed at the interface. Actually, voltage shifts are observed at hydrogen partial pressures as low as 10^{-10} torr (Dannetun *et al.*, 1984). It is also found that a small amount of oxygen introduces surface reaction sites that effectively empty the interface of hydrogen. Furthermore, oxygen behaves quite differently on a surface which is cleaned by Ar-sputtering in the vacuum chamber compared to a noncleaned surface (Petersson *et al.*, 1984b). For a noncleaned surface, oxygen appears to adsorb on the surface with a rather low heat of adsorption, oxygen comes readily off the surface also at low temperatures. On a cleaned surface, oxygen is strongly adsorbed and can at 100–150°C only be removed by hydrogen. This indicates the presence of two types of oxygen on the palladium surface. For a Pd gate gas sensor operating in air, hydrogen probably interacts with the first type of oxygen. The UHV-measurements have also yielded some interesting fundamental information about the reaction between hydrogen and oxygen on a clean Pd-surface (Petersson *et al.*, 1984a). An example of experimental results is given in Section VII,E. We believe that reaction studies and spectroscopic measurements in ultrahigh vacuum with techniques such as mass spectrometry, ultraviolet- and x-ray photoemission spectroscopy (UPS and XPS) and electron energy loss spectroscopy (EELS) will be very helpful in elucidating the behavior of Pd-gate gas sensors.

Another fundamental effect related both to the catalytic metal and to the insulator, is the spillover of hydrogen from the metal (Pd) to the insulator (SiO_2) surface outside the metal. This was observed as an increase in the depletion layer capacitance of Pd-SiO_2-Si capacitors on *p*-type silicon due to positive charges (protons) migrating out on the surface (Armgarth *et al.*, 1984). The positive charges created an extended inversion layer in the semiconductor, thereby increasing the semiconductor capacitance. Hydrogen spillover from a catalytic metal to its support is an important issue also in heterogeneous catalysis (Antonucci *et al.*, 1982).

E. INTERFACE OR BULK POTENTIAL CHANGES?

We have assumed without further explanation that the measured change in flat-band voltage ΔV is due to a dipole layer at the metal–insulator (or metal–semiconductor) interface,

$$\Delta V = \Delta V_{max} \theta_i$$

where θ_i is the coverage of hydrogen atoms at the interface. It is obvious that a change in the bulk Fermi level in the metal would also give rise to a

ΔV. There are, however, a number of reasons that exclude this as a main cause of the observed flat-band shift.

1. Langmuir Isotherms

It is found experimentally that

$$\frac{\Delta V / \Delta V_{max}}{1 - \Delta V / \Delta V_{max}} = f(\sqrt{P_{H_2}})$$

both in an inert atmosphere and in oxygen. This equation has the form of a Langmuir isotherm, which is common in surface adsorption and reactions. It stems from the fact that the number of surface (and interface) adsorption sites is limited. Furthermore, in pure argon, it was possible to determine the temperature dependence of the equilibrium constant k_0 in Eq. (25). It was found that the dissociation of hydrogen was exothermic, that is, that energy was gained when a hydrogen molecule dissociated into two hydrogen atoms. The energy gained, called the heat of adsorption, was about 1.1–1.2 eV per H_2 molecule. This is very close to the value found for surface adsorption sites for hydrogen on Pd, 1.1 eV/H_2 (Bond, 1974), whereas the adsorption energy for bulk hydrogen is much smaller, ~0.3 eV/H_2 (Lewis, 1967).

2. Difference in Surface and Interface Potential Changes

It is well known that adsorption of gases on a metal surface changes the surface potential and that the changes can be very large (~1 V) Hölzl and Schulte, 1979). The reason for the change in surface potential is that atoms or molecules give rise to dipoles when they adsorb. For a symmetric atom such as hydrogen the dipole is due to a displacement of the electron cloud either away from the metal (r-adsorption) or into the metal (s-adsorption) (Horiuti and Toya, 1969). The flat-band voltage changes observed correspond to s-adsorption at the interface. The proof of the surface origin of the potential changes is, however, not in the fact that the surface and interface potentials are similar for hydrogen in argon, but that they are different for a hydrogen–oxygen mixture. The Kelvin probe studies also indicated that, both at atmospheric pressure and in UHV, there are hydrogen adsorption sites at the interface that are not present on the surface.

The number of hydrogen atoms per bulk Pd atom is very low under typical experimental conditions, probably less than 0.001 (Lewis, 1967; Lundström et al., 1977). The expected change in bulk Fermi level is therefore extremely small, taking into account the density of states of the metal at the Fermi level. Appreciable amounts of hydrogen in the metal are found only under extreme conditions, such as low temperature, large hydrogen concentrations, and inert background atmosphere (see below).

3. Lack of Other Bulk Effects

A large change in the bulk Fermi level should also give rise to changes in, for example, the resistance of the thin Pd gate. We have made some studies that indicate that under normal conditions there are no resistance changes in the Pd film (Lundström and Söderberg, 1981–1982). At high hydrogen pressures in argon a bulk resistance change could be observed. This change was, however, connected with the filling of adsorption sites with smaller heat of adsorption than those involved in the normal hydrogen response in air.

4. Different Interfaces

We performed several experiments involving a modification of the Pd–SiO_2 interface. A thin but continuous layer of chromium at the interface destroys the hydrogen sensitivity; a thin but discontinuous layer changes only the magnitude of ΔV_{max}. A 10-nm layer of gold at the interface also destroys the gas sensitivity. These experiments indicate that not only are the catalytic activity of the metal surface and a high diffusion constant for hydrogen of importance, but the solubility of hydrogen atoms in the metal must be high, to enable them to reach the metal–insulator interface. Other modifications are related to the insulator. A Pd–NiO–SiO_2–Si structure shows a large number of slow interface adsorption sites, giving rise to hysteresis and long-term drift phenomena. A Pd–Al_2O_3–SiO_2–Si structure (see Fig. 18) has very few slow adsorption sites and shows no long-term drift due to hydrogen (Armgarth and Nylander, 1981; Dobos et al., 1984; Hua et al., 1984a).

5. Summary

The arguments above clearly demonstrate the interface origin of the flat-band voltage changes. A summary of our knowledge of the Pd–SiO_2 system is presented in a schematic energy diagram for hydrogen in Fig. 29. The different adsorption sites and phenomena denoted by the circled numbers are listed in the figure caption.

F. MODIFIED METAL GATES

Most studies to date have been made on "nonporous" Pd films, in which only hydrogen atoms reach the internal surface of the metal. The flat-band voltage change induced by a dipole layer is, however, a general

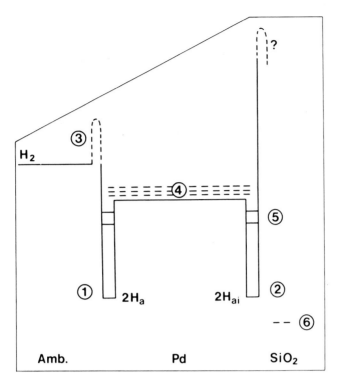

Fig. 29. Schematic summary of the adsorption sites for hydrogen and some other observed phenomena in the Pd–SiO$_2$ system. (1 and 2) "Normal" adsorption sites at the surface and interface, respectively, with similar properties. (3) An extra barrier for the dissociation of hydrogen, probably related to adsorbed oxygen atoms. (4) Bulk adsorption sites filled to any extent only in an inert atmosphere (at low temperatures and/or high hydrogen concentrations). (5) Hydrogen adsorption sites with small adsorption energies filled in experiments with hydrogen in an inert atmosphere and temperatures around 100–150°C. (6) Hydrogen introduced in the oxide probably interacting with (sodium) ions in the oxide (see Section VI).

phenomenon and is not connected specifically to hydrogen and palladium. For example, it has also been demonstrated for sodium in a mercury gate (Corker and Svensson, 1978).

A general type of gas sensor is obtained if the metal layer (or some other conductor) is made porous. In this case the metal grains become equipotential surfaces, whose potential is influenced by the adsorption of molecules that are dipoles or become polarized on adsorption. In such a case the difference between surface and interface potential disappears, as neatly demonstrated by Kelvin probe experiments (Fig. 30). In this case the porous film was made by oxidizing and reducing the Pd film on a

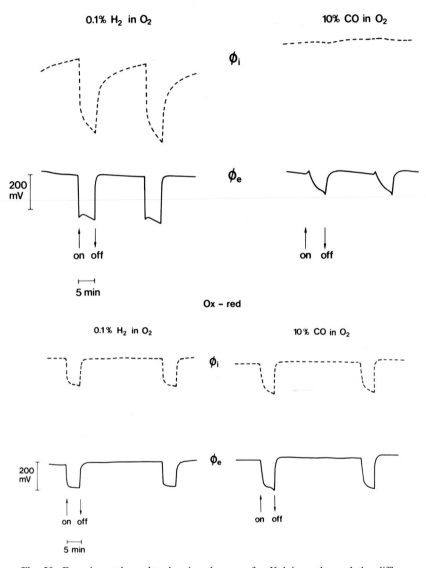

Fig. 30. Experimental results showing the use of a Kelvin probe and the difference between porous and nonporous Pd films (Söderberg *et al.*, 1980). "Evaporated" denotes a Pd device prepared in the normal way. "Ox–red" stands for a Pd film that was subjected to an oxidation–reduction step after preparation. ϕ_i (dashed line) indicates changes of the interface potential (or flat-band voltage); ϕ_e (solid line) indicates changes in the surface potential as measured with a Kelvin probe. Sample temperature was 150°C.

normal Pd–SiO$_2$–Si structure (Söderberg *et al.*, 1980). Other ways to make porous Pd films have also been tried (Dobos *et al.* 1980). In all cases the dipoles observed at the surface are also observed at the interface. Porous metal films therefore offer the possibility of developing general gas sensors, with reduced selectivity. We have observed sensitivity to CO, C$_2$H$_5$OH, CCl$_4$, etc. (Söderberg *et al.*, 1980; Lundström and Söderberg, 1981–1982). Dobos *et al.* (1980, 1983) and Krey *et al.* (1982–1983) reported on CO-sensitive porous Pd-gate transistors and Yamamoto *et al.* (1980) reported similar results for Pd–TiO$_2$ structures. Some selectivity can be obtained by operation at different temperatures since different molecules have different heats of adsorption. Interestingly, the porous gate should also work with noncatalytic conductors, the only requirement being that the adsorbed molecules give rise to surface potential changes.

Another modification is to use catalytic metals other than Pd. Yamamoto *et al.* (1980) tried a number of metals in metal–TiO$_2$ Schottky barriers and concluded that Pd was the superior metal. It has been shown that Pt works similarly to Pd in an inert atmosphere (Lundström and DiStefano, 1976) but that a large difference is observed in air or oxygen, palladium being much more sensitive to hydrogen than platinum (Armgarth *et al.* 1982a). This difference is due to the difference in oxygen adsorption and catalytic activity between the two metals. Dense Ni films exhibit no hydrogen response in air (at temperatures around 150°C), probably due to an inactive layer of nickel oxide. We also tried dense films of metals like Fe, Ru, and Rh as gates to obtain, for example, increased ammonia sensitivity. These experiments have so far been unsuccessful. Dobos *et al.* (1983) demonstrated CO sensitivity of porous Pt and SnO$_2$ gates.

Another approach is to change the catalytic properties of, for example, palladium through alloying with other metals or through surface modification to enhance certain reactions. We have evaporated Pt and Ir in thin layers on top of a palladium gate and observed an increase in the sensitivity to ammonia of a factor of 10–60, in the case of Ir accompanied by a decrease in hydrogen sensitivity (see Fig. 31). It was observed, however, that the thin metal layer must cover part of the oxide outside the Pd dot, indicating that part of the catalytic action takes place on the oxide (Spetz *et al.* 1983; Winquist *et al.*, 1983).

$C(V)$ measurements on structures with a thin catalytic layer (Pt) and a thick noncatalytic metal (Al) as a contact indicate that an ammonia-induced voltage shift is obtained at the thin metal-oxide interface. It is furthermore found that the important fact is that the catalytic metal is porous. For Pt maximum ammonia sensitivity is obtained for thicknesses around 20–30 nm (Spetz *et al.*, 1984). Apparently, reaction sites both on the insulator surface and on the catalytic metal are needed for an efficient

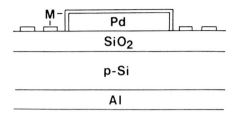

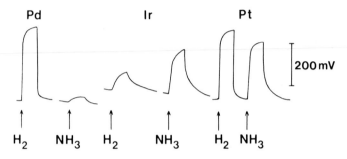

Fig. 31. H_2 and NH_3 sensitivity of modified Pd-MOS devices. A thin layer of a metal (M) was evaporated on top and outside the Pd spot. Gas concentrations: 100 ppm H_2 or NH_3 in 20% O_2 in argon as the carrier gas. Device temperature was 150°C. Gas pulses were 4 min long.

interaction with ammonia molecules. The dependence of the voltage shift on the ammonia concentration is similar in form to the hydrogen dependence, which suggests that ammonia molecules are decomposed on the surface and that water molecules are created in the presence of oxygen. It has been found that several thin catalytic metal films on SiO_2 show ammonia sensitivity, e.g., Pd, Pt, Ir and La.

Some other interesting developments have been made by Blackburn *et al.* (1983) who demonstrated the detection of dipolar molecules both in gases and in liquids by the use of a field effect transistor with a suspended metal mesh as a gate. Fluid samples (gases or liquids) can penetrate between the metal and the gate insulator. Polar molecules adsorbed either on the inside of the gate metal or on the insulator surface give rise to a threshold voltage shift in a manner similar to the hydrogen dipoles in the Pd-MOS devices. An example of methanol detection in toluene with a suspended Pt mesh is shown in Fig. 32. Stenberg and Dahlenbäck (1983) made an open-gate field effect transistor by etching away the oxide under

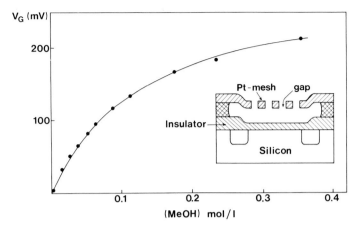

Fig. 32. Concentration–response curve of methanol in toluene for a suspended mesh Pt-gate transistor (from Blackburn *et al.*, 1983). The insert shows a schematic of the suspended mesh transistor.

a polysilicon gate at the source. This structure also becomes sensitive to polar molecules. For both types of devices, we expect that changes in the dielectric constant of the medium in the open-gate region will change the transistor parameters.

G. ROOM-TEMPERATURE OPERATION

The theoretical discussion in this section was focused on palladium sensors operated at elevated temperatures, the conditions for which a physical picture of their operation has started to emerge. The catalytic metal–(insulator)–semiconductor structures can, however, be used at room temperature, as pointed out earlier. There appear to be two main differences between room temperature and high-temperature operation (Lundström and Söderberg, 1981–1982): the former is limited by a smaller practical sensitivity and longer time constants. For nonporous films the exact mode of operation at low temperatures is not known; several phenomena may occur when the temperature is lowered. The deep hydrogen adsorption sites (with adsorption energies of ≈ 1 eV/H_2) will be saturated at extremely low hydrogen concentrations, especially since water adsorption on the Pd surface appears to block the reaction between hydrogen and oxygen. We therefore suggest that at room temperature other adsorption sites and even bulk hydrogen give rise to the observed voltage shifts.

For porous metal films the situation is even more complicated, as indicated by Yamamoto *et al.* (1980). At any rate, it is obvious that the reactions on the surface are slowed at lower temperatures. It has also been possible to operate the ammonia sensor with the thin extra catalytic metal layer at room temperature and still maintain a high ammonia sensitivity (Winquist *et al.*, 1984a,b). A careful comparison of low- and high-temperature operation of catalytic metal gate chemical sensors has still to be made.

VI. The Si–SiO₂ System, Properties of Relevance to Gas Sensors

A. BASIC PROPERTIES OF THE Si–SiO₂ SYSTEM

As mentioned earlier, the MOS technology is based on the unique properties of oxidized silicon. MOS devices are therefore very sensitive to the properties of the silicon dioxide film and the silicon–silicon dioxide interface. The MOS structure is fabricated by oxidation of single-crystal silicon. (100) silicon surfaces are normally preferred over (111) surfaces, but either may be used. Oxidation takes place under extremely clean conditions in oxygen, water vapor, or mixtures of the two. The oxidation temperature is normally 800–1200°C.

Sometimes additives like hydrogen chloride are used during oxidation. After oxidation the structure can be annealed at high temperatures in nitrogen or argon or at low temperatures in hydrogen or hydrogen–nitrogen mixtures. The low-temperature anneal is done at 400–500°C, often after metallization. After oxidation the structure is metallized by evaporation of the appropriate metal (normally aluminum); we have used a catalytic metal. It is important that the metallization be done under extremely clean conditions. Metallization is accomplished by evaporation in a vacuum, often by heating the metal with an electron gun.

Extreme care about cleanliness is necessary because the process is highly sensitive to impurities. Alkali metals such as sodium are particularly harmful. Even very small amounts of sodium (0.001 of a monolayer on the surface) may give rise to considerable drift in the flat-band voltage of the device, due to drift of ionized sodium inside the oxide layer.

The properties of the Si–SiO₂ system are determined by the defect structure in the system. The defect structure is controlled mainly by the oxide growth process, which is normally limited by diffusion of the oxidant (oxygen or water) through the oxide layer to the silicon. Intrinsic defects are formed as a result of the growth process and chemical impurities are introduced from the environment during the process.

Figure 33 illustrates the most important defects in the system (Deal, 1974): fixed oxide charge, surface traps, radiation-induced charge (or hole traps), and alkali ions. The fixed charge depends on the high-temperature oxidation process and is normally minimized by a short postoxidation anneal in nitrogen or argon. The surface trap density is normally mini-

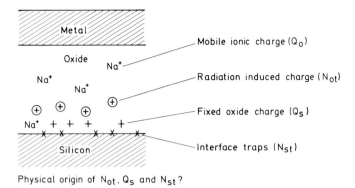

Physical origin of N_{ot}, Q_s and N_{st}?

Fig. 33. Electrical defects in the Si–SiO$_2$ system.

mized by a postoxidation low-temperature anneal in a hydrogen–nitrogen mixture. The nature of the fixed charge is not known, but it is believed to be related to excess silicon in the oxide close to the oxide–silicon interface. Surface traps are most probably caused by silicon dangling bonds at the interface (silicon surface atoms bonded to only three other atoms). Hydrogen treatment is assumed to give rise to hydrogen bonding to the silicon dangling bonds, thus making these defects electrically inactive.

The silicon oxide layer is normally sensitive to ionizing radiation (with energy larger than the oxide band gap, about 9 eV). Radiation creates electron–hole pairs, the holes of which are trapped in oxide defects, giving rise to an extra positive charge (DiMaria *et al.*, 1977). As a secondary effect new surface traps are created. Radiation damage can normally be annealed out at 300–500°C. Note that electron beam evaporation of a metal causes radiation damage of the oxide from soft X-rays created during the process. With this evaporation method it is necessary to use a postmetallization anneal.

Alkali ions, mainly sodium but also potassium, are detrimental to an MOS device because these ions—and hence, the ionic charge—will move around in the structure even at room temperature, causing considerable instability (Snow *et al.*, 1965). Normally the ions are loosely bound to the interfaces. Thus, in fabrication of these devices extreme care must be taken to avoid the incorporation of alkali ions, particularly during the oxidation and metallization processes. *In situ* cleaning of the furnace tubes can be used, as well as all other possible care. If the previously mentioned electron gun evaporation process of metallization is used, only the metal to be evaporated is heated—all other parts are water-cooled.

Even with the care mentioned above, ion contamination is hard to avoid. Therefore, methods for reducing the effect of ion contamination have been developed. On top of the oxide layer a thin phosphosilicate glass may be formed to trap the sodium ions, making them inactive. In a similar way a thin layer of silicon nitride may be deposited on the oxide. Such a layer hinders sodium diffusion because of its higher density. Finally, the oxide may be grown in the presence of chlorine compounds, such as hydrogen chloride. This leads to incorporation into the oxide–silicon interface of considerable amounts of chlorine, which has the ability to trap and neutralize moving sodium ions.

Chemical analysis of the oxide films often shows considerable amounts of inactive sodium (much more than of sodium ions). This inactive sodium is probably bonded to SiO^- groups. It is not clear whether it is of any importance.

It has been speculated that hydrogen ions also may move around in MOS structures; however, no clear evidence for hydrogen ion drift exists.

Finally, it may be of importance that the diffusivity of gases in silicon dioxide is rather high. Hydrogen, for example, moves quite freely in silicon dioxide films. In Table II we express the diffusivities as the temperatures at which appreciable diffusion of the actual gas takes place ($W/4\sqrt{Dt} = 1$ with $W = 100$ nm and $t = 30$ min; D is the diffusion constant). The table shows that water diffusion also may be of importance at low temperatures.

B.   STABILITY PROBLEMS

The stability of an MOS device is closely connected to the defects in the oxide and at the oxide–silicon interface. We can divide these defects into two groups: charges and traps.

Oxide and interface charges have a direct effect on the flat-band voltage [Eq. (11)] and the transistor threshold voltage [Eq. (12)]. Any change in the value of the charge or its position within the oxide will affect these

TABLE II

DIFFUSION OF GASES IN SILICON DIOXIDE

Defect	Diffusion temperature (°C)
He	−140
H_2	0
H_2O	100
O_2	300
N_2	300

voltages. A charge distribution $\rho(x)$ in the oxide ($x = 0$ at the metal) will contribute to the MOS flat-band voltage a value (Sze, 1981) of

$$- \frac{1}{\varepsilon_{ox}} \int_0^{W_{ox}} x\rho(x) \, dx \tag{37}$$

The most common charge instabilities are ion drift (mainly sodium ions), a negative bias stress effect, and radiation damage. Ion drift is normally observed in an MOS structure with positive bias to the gate. Normally, most ions are located close to the metal interface, thus keeping the charge contribution to the flat-band voltage very small [Eq. (37)]. Because of the positive gate voltage the ions drift toward the silicon, increasing the charge contribution to the flat-band voltage and making it more negative. This effect can be very large; a contaminated oxide may give shifts of 100 V. The amount of ion drift is normally tested in device fabrication by a so-called bias–temperature–stress method. The device is heated to 150°C, a negative bias is applied, the device is cooled with bias, and the flatband voltage is measured. This value corresponds to all sodium ions at the metal. The procedure is then repeated for positive bias and the flat-band voltage corresponding to all ions at the silicon is measured. The difference is a measure of the drifting sodium ions per unit area, according to

$$N_i = \frac{\varepsilon_{ox}}{q W_{ox}} \Delta V_i \tag{38}$$

where ΔV_i is the flat-band voltage difference.

A negative bias stress effect is observed during a negative bias to the metal electrode at an elevated temperature. During such treatment there is an increase in the positive charge at the silicon–silicon dioxide interface, making the flat-band voltage more negative. The phenomenon is not fully understood but is believed to be electrochemical in nature (Jeppson and Svensson, 1977).

Radiation damage also gives rise to a positive interface charge. This charge consists of photogenerated holes that are trapped at hole traps in the oxide. The nature of the hole traps is not known.

Interface or oxide traps may also give rise to instabilities. Interface traps are normally in equilibrium with the semiconductor, that is, their charge is given by the position of the semiconductor Fermi level at the interface. A fixed interface trap distribution will therefore not give rise to any instability because the charge will be a given function of the bias voltage (for times longer than the trap time constants of <1 ms). A varying interface trap density may, however, cause instabilities. Interface traps are known to be created in connection with both negative bias stress and radiation damage (Deal, 1974; Winokur and Sokoloski, 1976). Interface traps may disappear as a result of hydrogen anneal; if atomic hydrogen is available, an annealing temperature of 200°C may be effective (Johnson et al., 1980).

Oxide traps may be of importance for device stability because of their long time constants. If their occupation is changed, they may stay in the new charge state for very long times. Oxide traps may be charged by trapping radiation-induced holes and electrons, as mentioned above. Holes and electrons may also be injected into the oxide layer by other means, giving rise to trapping of charge carriers in the oxide traps. Electrons and holes may be injected by direct tunneling (from the semiconductor directly to an oxide trap) as a result of an applied voltage (slow trapping effect; Colbourne et al., 1974). A positive gate voltage will give rise to a slow positive drift of the flat-band voltage and vice versa. Finally, holes or electrons may be injected into the oxide from a backbiased pn junction in the semiconductor (e.g., the drain junction of an MOS transistor). In a backbiased pn junction the electric field may become very high, causing some electrons (holes) to be accelerated to an energy large enough to overcome the barrier into the oxide (Frohman-Bentchkowsky, 1974). The electron (hole) will then enter the oxide conduction (valence) band, from which it may be trapped in an oxide trap.

Most of the stability problems discussed above are not unique to chemical sensors but are properties of any MOS device. These problems may therefore be attacked in parallel with corresponding problems in ordinary MOS devices. There may, however, also exist stability problems that are unique to MOS chemical sensors. One example of such an effect could be the hydrogen-induced anneal of interface traps mentioned above. As (diatomic) hydrogen—and possibly also atomic hydrogen—is available in a Pd-MOS sensor, such an effect is plausible. Other interactions between hydrogen or other fast diffusers and the MOS structure may also be important. It appears, for example, that hydrogen also interacts with sodium ions in Pd-gate MOS structures, thus changing the sodium ion drift properties of the device (Svensson, 1980; Nylander et al., 1981, 1984).

Finally, we would like to comment on the metal–silicon dioxide interface, which has been investigated very little compared to the silicon dioxide–silicon interface. [Svensson (1980) has described some observations.] In the case of MOS gas sensors this interface is of crucial importance. The basic hydrogen sensitivity of the Pd-MOS device is a metal–silicon dioxide effect. Also, the hydrogen-induced drift (HID) is located at this interface (Nylander et al., 1981). Recent experiments indicate that the HID effect is caused by capture of H^+ by sodium-related defects in the oxide, close to the metal–oxide interface (Nylander et al., 1984). This interface thus plays an important role in device stability, and it is not surprising that modifications of this interface greatly influence device characteristics. A Pd–alumina (Al_2O_3) interface, for example, has shown a considerably reduced HID effect (Armgarth and Nylander, 1981; Dobos et al., 1984; Hua et al., 1984a). It is interesting to note that alumina has been used for similar reasons in ion-sensitive field effect devices (ISFETs) (Abe et al., 1979).

VII. Applications

Pd-MOS devices are now commercially available and a number of possible applications have been suggested, including:

1. A simple leak detector (Stiblert and Svensson, 1975; Armgarth et al., 1982b)
2. A smoke detector or fire alarm (Lundström et al., 1976), an oxygen monitor (Lundström and Söderberg, 1981)
3. A monitor for biochemical reactions (Danielsson et al., 1979; Winquist et al., 1982; Hörnsten et al., 1984; Berg et al., 1985; Cleland et al., 1984)
4. A detector of hydrogen in the process industry, for corrosion monitoring and for surveillance of battery charging

The ammonia sensitive MOS devices have found several biochemical uses (Winquist et al., 1984a,b). Some applications are briefly described below.

A. LEAK DETECTOR

A robust, light, small leak detector was developed with a Pd-MOS transistor as the sensing element (Stiblert and Svensson, 1975). This detector—the size of a "king size cigarette package"—was battery-driven and contained a temperature control circuit and a summing amplifier which was set to sound at a given hydrogen concentration. In a typical application the system to be tested was filled with 10% H_2 in N_2 (forming gas). The leak detector had an estimated performance close to that of currently available leak detectors. One big advantage is that hydrogen gas is both nontoxic and cheap.

Armgarth et al. (1982b) have developed a slightly larger (but still portable) leak detector with better performance and equipped with rechargeable batteries (Fig. 34) that can also be used as a (continuous) hydrogen monitor. It has been used successfully, for example, to track leaks in a buried gas-filled cable. Hydrogen monitors based on palladium–silicon dioxide–silicon Schottky diodes have also been constructed (Ito, 1981; Ito and Kojima, 1982). Ito reports very good long-term stability of these sensors when operated at low temperatures.

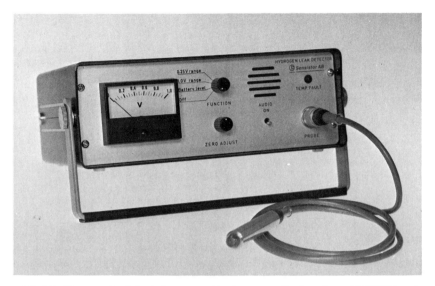

Fig. 34. Photograph of a leak detector and hydrogen monitor based on a Pd-MOS transistor with integrated heater and temperature sensor. (Courtesy of Sensistor AB, Sweden.)

B. FIRE ALARM

Before a fire starts the smoke from most materials contains an excess of hydrogen (or hydrogen-containing molecules) that can be detected by a Pd-MOS fire alarm (Lundström *et al.*, 1976). In this respect the Pd-MOS devices are interesting because they detect the fire "before it starts." Figure 35 shows the results of an experiment in which different materials were ignited in a hood. The amount of smoke detected can be estimated from the response to a cigarette, which was about 50 cm from and 40 cm below the detector. In order to avoid water vapor condensation and to increase the sensitivity and decrease the response time the devices were operated at an elevated temperature. Operation at or close to room temperature appears to be possible, especially with porous gate metals (Yamamoto *et al.*, 1980; Ito, 1981; Lundström and Söderberg, 1981–1982). It is observed, however, that both the speed of response and the sensitivity decrease with temperature.

C. OXYGEN MONITORING

A small amount of ambient hydrogen makes it possible to use Pd-MOS devices as sensors for oxygen—the reactions with oxygen on the metal surface determine the amount of hydrogen available at the internal

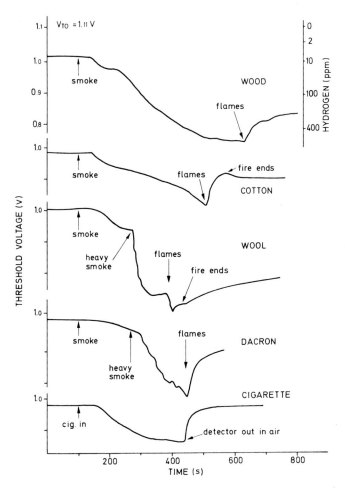

Fig. 35. Results of model experiments with the Pd-MOS device as a smoke detector (Lundström *et al.*, 1976). The right axis indicates the apparent amount of hydrogen in the smoke. Temperature of the sample was 150°C.

surface of the metal. More generally, all gases that react with hydrogen on the metal surface can be detected in the presence of hydrogen. No systematic study has yet been made of this potential application, although it has been observed that chlorine behaves similarily to oxygen. Furthermore, H_2O_2 and O_3 (ozone) act as strong oxidants for hydrogen on the Pd surface. H_2O_2 appears, for example, to empty hydrogen adsorption sites, which are not influenced in normal air (Lundström *et al.*, 1977).

Figure 28 is one example of the use of the Pd-MOS devices as oxygen monitors. We have found that with appropriate background concentrations of hydrogen, the devices are excellent and stable (Lundström and Söderberg, 1981). They can probably be developed into fast, reliable oxygen monitors. Possible applications could be as sensitive monitors for trace amounts of oxygen, for breath analysis and for the measurement of transcutaneous oxygen. In all cases a maximum differential sensitivity can be obtained through a proper choice of the background hydrogen concentration. The transition from a hydrogen-covered to an oxygen-covered surface mentioned in Section V,D can be utilized for accurate oxygen concentration limit detection.

D. BIOCHEMICAL REACTIONS AND MEDICAL DIAGNOSIS

There are many biochemical reactions that evolve a gas such as NH_3, H_2S, or H_2. A typical example is the decomposition of urea by the enzyme urease. Normally, the NH_3 created is measured in solution with special electrodes. It was found, however, that the ammonia in the gas above the sample cell could be conveniently monitored with an ammonia-sensitive Pd-MOS device (Danielsson et al., 1979). In these experiments no carrier gas was used; the device was applied above the surface of the solution.

The biochemical investigations have been extended to other systems. We have, for example, studied the hydrogen created in enzymatic reactions involving hydrogen dehydrogenase and NAD and NADH. Furthermore, microbial hydrogen production has been monitored with Pd-MOS devices (Winquist et al., 1982). It is interesting to note that hydrogen is produced in many fermentation processes and that a hydrogen monitor may have applications in food control (Cleland et al., 1984).

The high ammonia sensitivity of porous (thin) catalytic metal gates is, of course, very encouraging for biochemical applications. The devices produced so far appear to have a practical detection limit of about 1 ppm in air (Spetz et al., 1983; Winquist et al., 1983). Furthermore, they appear also to be useful for low-temperature applications. The calibration curves in Fig. 36 (Winquist et al., 1984b) were obtained with ammonium phosphate injected into a continuous stream of phosphate buffer, as indicated in the inset in Fig. 36. The ammonia in equilibrium with the ammonium ions was detected on the backside of a Teflon membrane by an Ir–Pd MOS kept slightly above room temperature. A detection limit of about 1 μM ammonium ions was found at pH 7.7. If short pulses (10–30 s) of ammonium phosphate were injected, the transient change of the flat-band voltage was linearly related to the ammonium ion concentration, since initially $d(\Delta V)/dt \sim [NH_3]$ in the air gap above the sensor. An enzyme probe has been developed using this ammonia sensor as indicated in Fig.

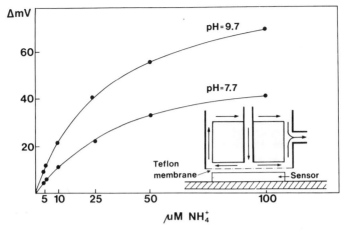

Fig. 36. Flat-band voltage shift of an Ir–Pd MOS structure caused by different concentrations of $(NH_4)_2PO_2$ in a phosphate buffer at different pH values. The buffer flowed over a Teflon membrane with the sensor placed on the other side of the membrane. The sensor was held at 35°C.

37, where also a calibration curve for urea detected by urease is shown (Winquist *et al.*, 1984a). The lowest detectable concentration was 0.01 mM of urea. This application can be extended to other enzyme–substrate systems. The sensor can of course also be mounted in a probe behind a gas permeable membrane.

We believe that the biochemical applications of gas sensors will enable

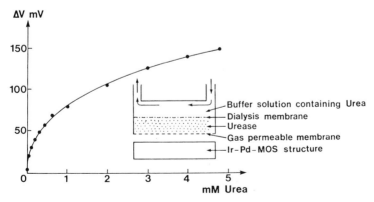

Fig. 37. Schematic drawing of an enzyme probe based on an ammonia sensitive Ir-Pd MOS structure. The curve shows the observed steady response to urea at different concentrations. Steady state was obtained about 3 min after a change in urea concentration. Sensor temperature 35°C.

continuous monitoring of parameters and processes that have previously been difficult to study. The combination of a (bio)chemical reaction and a gas sensor extends the uses of the sensor. Such a combination can be used, for example, as a monitor for substances that destroy the activity of enzymes (e.g., heavy metals). If a small amount of hydrogen was inhaled its disappearance from the lungs could be observed through the use of a Pd-gate device placed in the exhaled gas, thus providing a new method for studying ventilation. In certain diseases or malfunctions hydrogen or ammonia is produced in excess. Breath analysis could be a new technique for diagnosis of certain illnesses. An evaluation of the hydrogen monitor in Fig. 34 for the investigation of lactose malabsorption is under way; the results so far are promising (Berg et al., 1985). A typical result is shown in Fig. 38.

E. FURTHER APPLICATIONS

We have tried to combine the Pd-MOS sensor with other catalytic reactions to develop monitors for hydrocarbons and so on. One difficulty

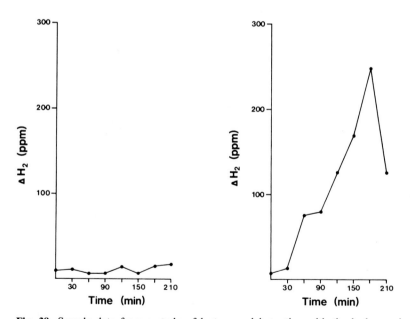

Fig. 38. Sample data from a study of lactose malabsorption with the hydrogen leak detector in Fig. 34. The curve on the left is the hydrogen content in breath for a normal person; the curve on the right shows hydrogen content for a person with lactose malabsorption. Time on the abscissa is time after intake of lactose. Hydrogen concentrations were determined by comparison with a calibration curve for the leak detector.

is that most commercial catalysts normally do not operate in air. Some preliminary experiments indicated, however, that a hot catalyst slug in front of the Pd-MOS device can create enough hydrogen even in a background of oxygen. The best result so far was obtained with an MgO/SiO_2 catalyst at a temperature around 600°C. Poteat and Lalevic (1981) demonstrated that Pd-MOS structures (at room temperature) react to high concentrations of hydrocarbons.

Other possible applications include corrosion monitoring, measurements of H_2, NH_3, and H_2S in the process industry, and, of course, scientific applications. These include the study of catalytic reactions *per se* (Petersson *et al.*, 1982a,b, 1984a), but also of hydrogen production in photoelectrochemical reactions and by microorganisms.

A result of fundamental importance for the understanding of catalytic reactions on metal surfaces is illustrated in Fig. 39. A Pd–SiO₂–Si structure was subjected to oxygen in a UHV chamber. Adsorption of oxygen and desorption of hydrogen were followed by a Kelvin probe (ϕ_e) and the $C(V)$ shift (ϕ_i) of the Pd-MOS structure, respectively. Oxygen was removed from the chamber (at "off" in Fig. 39) and, after a certain time, hydrogen was let into the chamber (at "on" in Fig. 39). Water production

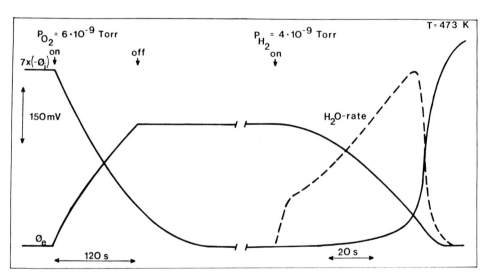

Fig. 39. Catalytic reactions in UHV studied with a Pd-MOS structure. The surface potential (ϕ_e) of the Pd film was measured with a Kelvin probe. Changes in the potential at the Pd–oxide interface (ϕ_i) were monitored as shifts in the $C(V)$ curve (i.e., the "normal" ΔV). Note that the ϕ_i curve is expanded seven times. Water production on the Pd surface was measured with a mass spectrometer. See text for further description. (From Petersson *et al.*, 1984a.)

on the Pd surface was monitored with a mass spectrometer (dashed curve in Fig. 39). It was observed that hydrogen reaches the Pd–SiO$_2$ interface (i.e., ϕ_i changes rapidly) first when almost all oxygen atoms on the surface are consumed. This means that hydrogen must have a large lateral mobility on Pd surfaces (Petersson *et al.*, 1984a).

It is interesting to note that Pd-gate devices can also be used at low temperatures. We have immersed a Pd-MOS device in an electrolyte and have been able to monitor the amount of hydrogen dissolved in the electrolyte (Lundström *et al.*, 1983). Preliminary studies indicate that it may be possible to use Pd gates as sensors for both pH and pH$_2$ with the use of a second electrode in the electrolyte.

We discussed briefly the use of hydrogen sensitivity in an indirect way to detect oxygen. Other gases that react with hydrogen on the catalytic metal surface can also be detected in that way—for example, O$_3$, H$_2$O$_2$, Cl$_2$, and others. Furthermore, a certain selectivity can be obtained through the use of (polymeric) membranes in front of the metal gate.

We mentioned that it is possible to change the sensitivity to ammonia through modification of the metal gate. There are reasons to believe that other gases can be "specifically" detected through other modifications. Multisensor chips for continuous monitoring of a gas mixture are therefore a possibility in the near future. A currently realizable technique is to use unmodified sensors at different temperatures, as many catalytic reactions are temperature-dependent. with different activation energies. Porous gate metals offer another interesting possibility for extending the use of metal gate gas sensors to other gases.

VIII. Conclusions

Pd-MOS and similar devices belong to the same class as ISFETs. These devices are of great interest because they combine silicon technology with specific chemical sensitivity. There is every reason to believe that stable and reliable chemical sensors will result from the development of both gas- and ion-sensitive silicon devices. There are also some problems shared by gas- and ion-sensitive devices, namely stability and selectivity. Stability is related to the physics of silicon devices and can be improved with better fabrication methods and new types of device structures. Selectivity depends on the "intelligent" sensing layer used as the gate of the devices. Further research will undoubtedly provide innovative solutions to these problems.

Probably the most important advantage of silicon-based gas sensors over other solid state gas sensors is that they can be fabricated using conventional integrated circuit technology, which permits reproducible mass production and integration of control functions on the same chip. This chapter summarizes our present knowledge of one type of silicon sensor, which has a high selectivity for one chemical species. We have also shown how this selectivity can be modified or destroyed, leading to new ways to develop sensors based on changes in the surface field of a semiconductor due to adsorbed (gas) molecules.

ACKNOWLEDGMENTS

We are grateful to our colleagues and students at the Research Laboratory of Electronics, Gothenburg, and at the Department of Physics and Measurement Technology, Linköping, who have contributed to the work described in this chapter.

Our research on gas sensors is supported by grants from the National Swedish Board for Technical Development and our research on catalytic reactions by the Swedish Natural Science Research Council.

REFERENCES

Abe, H., Esashi, M., and Matsuo, T. (1979). *IEEE Trans. Electron Devices* **ED-26,** 1939–1944.
Antonucci, P., van Truong, N., Giordano, N., and Maggiore, R. (1982). *J. Catal.* **75,** 140–150.
Armgarth, M., and Nylander, C. (1981). *Appl. Phys. Lett.* **39,** 91–92.
Armgarth, M., and Nylander, C. (1982). *IEEE Electron Device Lett.* **EDL-3,** 384–386.
Armgarth, M., Söderberg, D., and Lundström, I. (1982a). *Appl. Phys. Lett.* **41,** 654–655.
Armgarth, M., Nylander, C., Sundgren, H., and Lundström, I. (1982b). *Proc. World Hydrogen Energy Conf., 4th, 1982,* pp. 1717–1723.
Armgarth, M., Nylander, C., Svensson, C., and Lundström, I. (1984). *J. Appl. Phys.* **56,** 2956–2963.
Berg, A., Eriksson, M., Barany, F., Einarsson, K., Sundgren, H., Nylander, C., Lundström, I., and Blomstrand, R. (1985). *Scand. J. Gastroenterol.* (in press).
Blackburn, G. F., Levy, M., and Janata, J. (1983). *Appl. Phys. Lett.* **47,** 700–701.
Bond, G. C. (1974). "Heterogeneous Catalysis: Principles and Applications." Oxford Univ. Press (Clarendon), London and New York.
Cleland, N., Hörnsten, E. G., Elwing, H., Enfors, S.-O., and Lundström, I. (1984). *Appl. Microbiol. Biotech.* **20,** 268–270.
Colbourne, E. D., Coverley, G. P., and Behara, S. K. (1974). *Proc. IEEE* **62,** 244–259.
Corker, G. A., and Svensson, C. M. (1978). *J. Electrochem. Soc.* **125,** 1881–1883.
Danielsson, B., Lundström, I., Mosbach, K., and Stiblert, L. (1979). *Anal. Lett.* **12,** 1189–1199.
Dannetun, H. M., Petersson, L. G., Söderberg, D., and Lundström, I. (1984). *Appl. Surf. Sci.* **17,** 259–264.
Deal, B. E. (1974). *J. Electrochem. Soc.* **121,** 198C–205C.
DiMaria, D. J., Weinberg, Z. A., and Aitken, J. M. (1977). *J. Appl. Phys.* **48,** 898–906.

Dobos, K., Hoefflinger, B., and Zimmer, G. (1980). *Proc. Int. Vac. Congr., 8th, 1980,* Vol. I, pp. 743–745.

Dobos, K., Krey, D., and Zimmer, G. (1983). *Proc. Int. Meet. Chem. Sens., 1983,* pp. 464–467.

Dobos, K., Armgarth, M., Zimmer, G., and Lundström, I. (1984). *IEEE Trans. Electron. Dev.* **ED-31,** 508–510.

Fare, T., and Zemel, J. (1984). Personal communication.

Fonash, S. J., Huston, H., and Ashok, S. (1982). *Sens. Actuators* **2,** 363–369.

Froman-Bentchkowsky, D. (1974). *Solid-State Electron.* **17,** 517–529.

Green, M. A., King, F. D., and Shewchun, J. (1974). *Solid-State Electron.* **17,** 551–561.

Harris, L. (1980). *J. Electrochem. Soc.* **127,** 2657–2662.

Hölzl, J., and Schulte, F. K. (1979). "Solid Surface Physics." Springer-Verlag, Berlin and New York.

Horiuti, J., and Toya, T. (1969). *In* "Solid State Surface Science" (M. Green, ed.), Vol. 1, pp. 1–86. Dekker, New York.

Hörnsten, E. G., Elwing, H., Kihlström, E., and Lundström, I. (1985). *J. Antimicrobial Chemotherapy* (in press).

Hua, T. H., Armgarth, M., and Lundström, I. (1984a). *Proc. 11th Nord. Semiconductor Meet., 1984,* pp. 229–232.

Hua, T. H., Armgarth, M., and Lundström, I. (1984b). To be published.

Ito, K. (1979). *Surf. Sci.* **86,** 345–352.

Ito, K. (1981). *Jpn. J. Appl. Phys.* **20,** L753–L756.

Ito, K., and Kojima, K. (1982). *Int. J. Hydrogen Energy* **7,** 495–497.

Jeppson, K. O., and Svensson, C. M. (1977). *J. Appl. Phys.* **48,** 2004–2014.

Johnson, N. M., Biegelsen, D. K., and Moyer, M. D. (1980). *In* "The Physics of MOS Insulators" (G. Lucovsky, S. T. Pantelides, and F. L. Galeener, eds.), pp. 311–315. Pergamon, Oxford.

Kawamura, K., and Yamamoto, T. (1983a). *Proc. Int. Meet. Chem. Sens., 1983,* pp. 459–463.

Kawamura, K., and Yamamoto, T. (1983b). *IEEE Electron Device Lett.* **EDL-4,** 88–89.

Keramati, B., and Zemel, J. N. (1978). *In* "The Physics of SiO_2 and its Interfaces" (S. T. Pantelides, ed.), pp. 459–463. Pergamon, Oxford.

Keramati, B., and Zemel, J. N. (1982a). *J. Appl. Phys.* **53,** 1091–1099.

Keramati, B., and Zemel, J. N. (1982b). *J. Appl. Phys.* **53,** 1100–1109.

Krey, D., Dobos, K., and Zimmer, G. (1982–1983). *Sens. Actuators* **3,** 169–177.

Kurtin, S., McGill, T. C., and Mead, C. A. (1969). *Phys. Rev. Lett.* **22,** 1433–1436.

Lewis, F. A. (1967). "The Palladium Hydrogen System." Academic Press, London.

Lundström, I. (1981). *Sens. Actuators* **1,** 403–426.

Lundström, I., and DiStefano, T. (1976). *Solid State Commun.* **19,** 871–875.

Lundström, I., and Söderberg, D. (1981). *In* "Monitoring of Vital Parameters during Extracorporeal Circulation" (H. P. Kimmich, ed.), pp. 291–296. Karger, Basel.

Lundström, I., and Söderberg, D. (1981–1982). *Sens. Actuators.* **2,** 105–138.

Lundström, I., and Söderberg, D. (1982). *Appl. Surf. Sci.* **10,** 506–522.

Lundström, I., Shivaraman, M. S., Svensson, C., and Lundkvist, L. (1975a). *Appl. Phys. Lett.* **26,** 55–57.

Lundström, I., Shivaraman, M. S., and Svensson, C. (1975b). *J. Appl. Phys.* **46,** 3876–3881.

Lundström, I., Shivaraman, M. S., Stiblert, L., and Svensson, C. (1976). *Rev. Sci. Instrum.* **47,** 738–740.

Lundström, I., Shivaraman, M. S., Svensson, C. (1977). *Surf. Sci.* **64,** 497–519.

Lundström, I., Nylander, C., and Spetz, A. (1983). *Electron. Lett.* **19,** 249–251.

Nylander, C., Armgarth, M., and Svensson, C. (1981). *In* "Insulating Films on Semiconduc-

tors" (M. Schulz and G. Pensl, eds.), pp. 195–198. Springer-Verlag, Berlin and New York.
Nylander, C., Armgarth, M., and Svensson, C. (1984). *J. Appl. Phys.* **56**, 1177–1188.
Petersson, L.-G., Dannetun, H., Karlsson, S.-E., and Lundström, I. (1982a). *Phys. Scr.* **25**, 818–825.
Petersson, L.-G., Dannetun, H., Karlsson, S.-E., and Lundström, I. (1982b). *Surf. Sci.* **117**, 676–684.
Petersson, L.-G., Dannetun, H., and Lundström, I. (1984a). *Phys. Rev. Lett.* **52**, 1806–1809.
Petersson, L.-G., Dannetun, H., Fogelberg, J., and Lundström, I. (1984b). To be published.
Plihal, M. (1977). *Siemens Forsch. Entwicklungsber.* **6**, 63–59.
Poteat, T. L., and Lalevic, B. (1981). *IEEE Electron Device Lett.* **EDL-2**, 82–84.
Poteat, T. L., and Lalevic, B. (1982). *IEEE Trans. Electron Devices* **ED-29**, 123–129.
Poteat, T. L., Lalevic, B., Kuliyev, B., Yousuf, M., and Chen, M. (1983). *J. Electron. Mater.* **12**, 181–214.
Ruths, P. F., Ashok, S., Fonash, S. J., and Ruths, J. M. (1981). *IEEE Trans. Electron Devices* **ED-28**, 1003–1009.
Shivaraman, M. S. (1976). *J. Appl. Phys.* **47**, 3592–3593.
Shivaraman, M. S., and Svensson, C. (1976). *J. Electrochem. Soc.* **123**, 1258.
Shivaraman, M. S., Lundström, I., Svensson, C., and Hammarsten, H. (1976). *Electron. Lett.* **12**, 483–484.
Snow, E. H., Grove, A. S., Deal, B. E., and Sah, C. T. (1965). *J. Appl. Phys.* **36**, 1664–1673.
Söderberg, D. (1983). Ph.D. Thesis, Linköping University.
Söderberg, D., and Lundström, I. (1980). *Solid State Commun.* **35**, 169–174.
Söderberg, D., and Lundström, I. (1983). *Solid State Commun.* **45**, 431–434.
Söderberg, D., Lundström, I., and Svensson, C. (1980). *Mater. Sci. Eng.* **42**, 141–144.
Spetz, A., Winquist, F., Nylander, C., and Lundström, I. (1983). *Proc. Int. Meet. Chem. Sens., 1983*, pp. 479–487.
Spetz, A., Armgarth, M., and Lundström, I. (1984). *Proc. 11th Nord. Semiconductor Meet. 1984*, pp. 233–236.
Steele, M. C., and MacIver, B. A. (1976). *Appl. Phys. Lett.* **28**, 687–688.
Steele, M. C., Hile, J. W., and MacIver, B. A. (1976). *J. Appl. Phys.* **47**, 2537–2538.
Stenberg, M., and Dahlenback, B. I. (1983). *Sens. Actuators* **4**, 273–281.
Stiblert, L., and Svensson, C. (1975). *Rev. Sci. Instrum.* **46**, 1206–1208.
Svensson, C. (1980). *In* "The Physics of MOS Insulators" (G. Lucovsky, S. T. Pantelides, and F. L. Galeener, eds.), pp. 260–264. Pergamon, Oxford.
Sze, S. M. (1981). "Physics of Semiconductor Devices." Wiley (Interscience), New York.
Winokur, P. S., and Sokoloski, M. M. (1976). *Appl. Phys. Lett.* **28**, 627–630.
Winquist, F., Danielsson, B., Lundström, I., and Mosbach, K. (1982). *Appl. Biochem. Biotechnol.* **7**, 135–139.
Winquist, F., Spetz, A., Armgarth, M., Nylander, C., and Lundström, I. (1983). *Appl. Phys. Lett.* **43**, 839–841.
Winquist, F., Spetz, A., Danielsson, B., and Lundström, I. (1984a). *Anal. Chim. Acta.* **163**, 143–149.
Winquist, F., Spetz, A., Lundström, I., and Danielsson, B. (1984b). *Anal. Chim. Acta.* **164**, 127–138.
Yamamoto, N., Tonomura, S., Matsuoka, T., and Tsubomura, H. (1980). *Surf. Sci.* **92**, 400–406.
Yamamoto, N., Tonomura, S., Matsuoka, T., and Tsubomura, H. (1981). *J. Appl. Phys.* **52**, 6227–6230.
Yang, E. S. (1978). "Fundamentals of Semiconductor Devices." McGraw-Hill, New York.
Yousuf, M., Kuliyev, B., Lalevic, B., and Poteat, T. L. (1982). *Solid-State Electron.* **25**, 753–758.

2

Chemically Sensitive Field Effect Transistors

JIŘÍ JANATA

DEPARTMENT OF BIOENGINEERING
THE UNIVERSITY OF UTAH
SALT LAKE CITY, UTAH

I. General Theory

A. HISTORICAL PERSPECTIVE

Developments in electronics and computer technologies over the past 40 years have had a major impact on analytical chemistry: operation amplifiers revolutionized almost every area of instrumental analysis in the 1950s and 1960s and, more recently, microprocessors have improved the precision and convenience of use of existing analytical techniques far beyond previous limits. Lasers and dedicated digital circuits have led to completely new methods of instrumental analysis. The role that electronics plays can be clearly traced in the history of the old electroanalytical technique—potentiometry. The fundamental concept of potentiometry is simple: one measures the difference of inner potentials of the metal of the indicator electrode and a well-defined reference electrode. However, in reality the electrical properties of the materials, nonidealities of the reference electrode, and impedance characteristics of the measuring circuit must all be considered simultaneously. When J. C. Poggendorf used his potential compensation technique in the middle of the nineteenth century, it was limited to low-resistance metallic electrodes. Furthermore, his technique was discrete, therefore totally unsuitable for continuous monitoring. The invention of tube voltmeters at the beginning of this century started the era of modern potentiometry. The relatively high impedance of tube voltmeters allowed the use of high-resistance electrodes, namely glass electrodes, and the signal could be monitored continuously. While the introduction of solid state voltmeters in the 1950s has not substantially expanded the scope of potentiometry, it has improved the precision and convenience of use of the electrodes. Furthermore, it coincided with the growth of the many types of ion-selective electrodes that are still in use today. The input resistance of a modern electrometer is in the range of 10^8 to 10^{15} ohms with a typical input current of 10^{-9} A, which is adequately low for modern ion-selective electrodes.

Ion-sensitive field effect transistors (ISFETs), introduced in the early 1970s (Bergveld, 1970; Matsuo *et al.*, 1971), represent another qualitative step in the evolution of potentiometry by integrating a chemically sensitive membrane with solid state electronics. They belong to the broad category of chemically sensitive solid state devices (CSSDs) (Zemel, 1975). It will be shown in this chapter that new types of electroanalytical probes sensitive to ions can be developed as the result of this integration.

B. QUALITATIVE COMPARISON OF ISE, CWE, AND ISFET

The similarity between the ion-selective electrode (ISE) and the ISFET is a good starting point in the discussion of the principle of operation of these devices. The relationship is shown graphically in Fig. 1, in which the conventional ISE and a reference electrode R are shown connected to the insulated gate field effect transistor (IGFET) input of a voltmeter (a). In the next diagram (b) the lead connecting the ISE is shortened. Finally, the ion-selective membrane M is attached directly to the insulator of the input transistor; at this stage the integration process is completed (c,d). There are, however, significant differences between the two devices: first, in the conventional ISE the ion-selective membrane is placed between the sample and the reference solution. The potential profile in this case is symmetrical (Fig. 2). This is not so with the ISFET; in this case, the membrane has the sample solution on one side but on the other it is in contact with a solid phase, which can be either conducting or

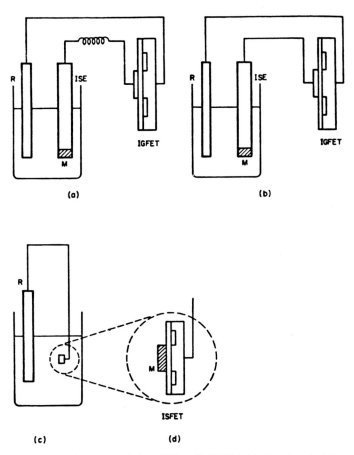

Fig. 1. Evolution of ISFET (c,d) from ISE and IGFET (a,b). (Reprinted with permission from Janata and Huber, 1980.)

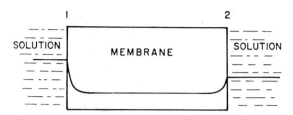

Fig. 2. Symmetrical arrangement of ISE membrane.

insulating, and, furthermore, this interface can be either capacitive or resistive. From the practical point of view, this means that any nonideal processes that may take place at the surface of the membrane, such as hydrolytic degradation of the membrane, will not cancel out in the ISFET. This may lead to long-term drift in these devices, which is a difficult and unavoidable problem.

Another question arises about the nature of the membrane–gate insulator interface. If the membrane is deposited directly over the insulator, there may be at this interface a charge distribution that corresponds to the potential profile shown in Fig. 3. This profile should become stable after the initial period of hydration. Any change in the solution composition is

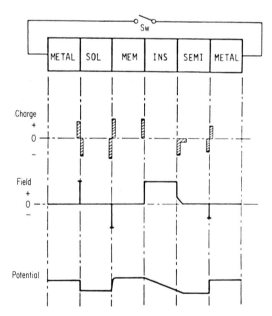

Fig. 3. Charge, field, and potential profiles of the ISFET gate. (Reprinted with permission from Janata and Huber, 1980.)

unlikely to result in a redistribution of charge if the membrane is thick (~ 100 μm) and ion-selective, although it cannot be ruled out *a priori*. In reality, all insulators used in ISFETs exhibit some pH sensitivity. When an ion-selective membrane is applied over this pH-sensitive "insulator" it means, in effect, that another "ion-selective layer" has been interposed between this membrane and the true insulator. However, since the insulator does not hydrate completely through, there is always a truly polarized interface at the end of this structure. It is, therefore, necessary to keep the applied electrical field constant so as not to disturb the electrostatic charge distribution at that interface. That can be accomplished relatively easily by operating the device at a constant gate charge (constant drain current), as will be shown later.

Somewhat less uncertainty about the structure of the interface exists when an electrically floating layer of metal is placed between the top insulator and the ion-selective membrane. Such devices have been made and some of them have shown a marked improvement in their dynamic response (Smith *et al.*, 1980). Again, the membrane–metal interface can be either polarized (capacitive) or nonpolarized (resistive). The nonpolarized case is, in principle, identical to the one discussed in the previous section. The potential distribution through an ISFET structure with a polarized membrane–metal interface is shown in Fig. 4.

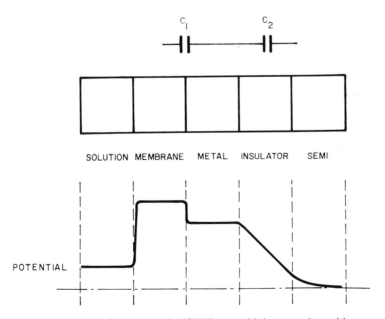

Fig. 4. Potential profile through the ISFET gate with interposed metal layer.

Let us represent the interfacial capacitance of this interface by equivalent capacitance C_1, while C_2 is the capacitance of the metal–insulator–semiconductor gate. Since there are no other capacitances involved, any change ΔV in the difference between the inner potentials of the membrane and the semiconductor caused either by the change of applied potential or by the change of the Nernst potential at the membrane–solution interface will be distributed across this capacitive divider according to Eq. (1)

$$C_1 \Delta V_1 = C_2 \Delta V_2 \tag{1}$$

If the externally applied potential is used to compensate for the chemically generated change of potential, this relationship will be constant and the capacitive nature of the membrane–metal interface will be of no consequence to the operation of the device.

It should be noted, however, that we have not specified the thickness of the metal layer; it can be, for example, 2000 Å, as it is in the case of ISFETs with a floating metal gate (Smith *et al.*, 1980), or it can be tens of centimeters of lead wire, as it typically is with coated wire electrodes that are then connected to, for example, an FET input of a voltmeter. Theoretically, there is no difference, because the thermodynamics of the interphase relationships does not depend on the thickness of the individual layers. Practically, the difference can be great; the surface area of the lead is one plate of a parasitic capacitance C_p that has poorly defined and variable value. There are also potentially large, poorly defined leakage resistances R_p to the shielding input gate protection diodes, and so on. The capacitive divider is then

$$C_1 V_1 = [C_2 + C_p(t)] V_2 \tag{2}$$

where $C_p(t)$ is a time-variable parasitic capacitance. Consequently, the gate voltage V_2 will vary in an unpredictable manner with $C_p(t)$. Furthermore, the parasitic resistance R_p will superimpose a variable time constant $C_p R_p$ on any change of voltage distribution. In practical terms, such a structure will exhibit unstable behavior. On the other hand, there is no problem with the nonpolarized (resistive) interface because the inner potential of the metal is always uniquely related to the inner potential of the membrane by the charge-transferring species i. Correctly, a number of researchers (Fjeldy and Nagy, 1980; Buck and Hackleman, 1977) designed the solid state internal contacts in such a way that each interface is nonpolarized. One such device is shown in Fig. 5. Of course, if an FET preamplifier is used the last interface is always polarized. This argument can be rephrased as follows: if there is a capacitor in series with the FET gate capacitor, it must be invariable. This condition is relatively easily met in CHEMFETs or in some hybrid structures because the geometry of

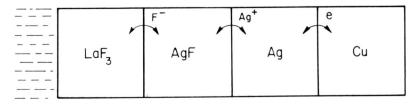

Fig. 5. Reversible solid state contact configuration.

the lead and its immediate surrounding is fixed; however, it can lead to severe complications with coated wire electrodes with a polarized or semipolarized membrane–metal interface. In very practical terms it can be stated that in the case of a capacitive membrane–metal interface, the problems increase with the length of the connecting lead. Nevertheless, ISFETs, hybrid structures, and coated wire electrodes are all the same in principle, the difference being only practical.

II. Thermodynamics of ISFETs

In the ISFET (Fig. 6) the metal gate is replaced with a reference electrode, a solution, and an ion-sensitive membrane. The rest of the device is protected by a suitable encapsulant. The heart of the ISFET is the gate. It was shown in the preceding chapter how the gate voltage V_G controls the drain current I_D in the transistor drain current equations. In order to describe quantitatively the mechanism of operation of the ISFET it is necessary to perform a thermodynamic analysis of the structure shown in Fig. 7, which represents a cross section through the whole measuring circuit, including the reference electrode and the connecting leads. This figure illustrates the simplest case, in which the reference electrode (1) is described by the equilibrium condition

$$M \rightleftarrows M^+ + e^-$$

A typical example of such an electrode would be a silver wire immersed in solution of silver ions. Let us further assume that the solution (2) also contains a small quantity of ions that can permeate reversibly into and out of the membrane (3), which therefore forms a nonpolarized interface. A possible example would be a solution (2) containing $0.01\ M$ $AgNO_3$ and $1 \times 10^{-4}\ M$ KNO_3 with a potassium-ion-sensitive membrane (3), such as used in the equivalent ISE. The insulator (4) is assumed to be ideal, that is, no charge can cross it, and it is thicker than the electron tunneling distance ($d \geq 100\ \text{Å}$). Layer (5) is the transistor semiconductor (such as p-silicon). Metal (6) will be identical with metal (1). A switch, SW, represents operation with (SW closed) and without (SW open) a reference

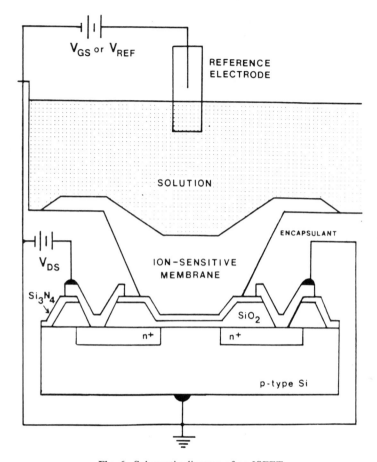

Fig. 6. Schematic diagram of an ISFET.

electrode. The charge, field, and potential profiles across this structure are also shown in Fig. 7. Note that this is a very simplified case. A liquid junction of the reference electrode, a dual-layer insulator, trapped charges in the insulator, surface states at the insulator–semiconductor interface, channel doping, and a multitude of connecting metals have been omitted for simplicity. Similar charge, field, and potential profiles taking into account some of these elements have been published by Sze (1969).

From the thermodynamic point of view, this is a multiphase system for which the Gibbs equation must apply at equilibrium at each interface:

$$\sum_i dn_1 \, \bar{\mu}_i = 0 \qquad (3)$$

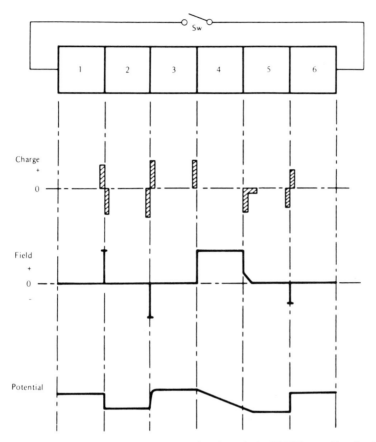

Fig. 7. Charge, field, and potential distribution through the ISFET gate. (Reprinted with permission from Janata and Huber, 1980.)

where dn_i is the number of species i transported across individual interfaces and $\bar{\mu}_i$ is the electrochemical potential of species i. Equation (3) can be expanded to yield the relationship

$$\bar{\mu}_3 \equiv \bar{\mu}_2 \equiv \bar{\mu}_1 \equiv \bar{\mu}_6 \equiv \bar{\mu}_5 \tag{4}$$

which simply expresses the fact that in the case of an ideal insulator (4) the only way this system can reach equilibrium is through the external pathway (1)–(6). If the switch SW is open, equivalent to operation without a reference electrode (or a signal return), then

$$\bar{\mu}_1 \neq \bar{\mu}_6 \tag{5}$$

The inequality of Fermi levels in metals (1) and (6) results in inequality of $\bar{\mu}_3$ and $\bar{\mu}_5$. Thus, the basic condition for stable operation of the ISFET is not satisfied.

Let us now analyze the circuit in Fig. 7. The inner potential in the semiconductor (5) can be expressed as

$$\phi_5 = \frac{1}{F}(\mu_5^e - \bar{\mu}_5^e) \tag{6}$$

where μ_5^e is the chemical potential of an electron in the semiconductor (the electron–lattice interaction energy) and $\bar{\mu}_5^e$ is the electrochemical potential of an electron in phase 5, normally known as the Fermi level. Similarly, for membrane (3) the inner potential ϕ_3 is

$$\phi_3 = \frac{1}{z^i F}(\bar{\mu}_3^i - \mu_3^i) \tag{7}$$

where z^i, the number of elementary charges, is positive for cations and negative for anions, and $\bar{\mu}_3^i$ and μ_3^i are the electrochemical and chemical potentials of species i in phase 3, respectively. The potential difference across the insulator and the semiconductor space charge region is then

$$\phi_5 - \phi_3 = \frac{1}{F}(\mu_5^e - \bar{\mu}_5^e) - \frac{1}{z^i F}(\bar{\mu}_3^i - \mu_3^i) \tag{8}$$

It is now essential to identify the relationship between species i in the membrane (3) and electrons in the semiconductor (5).

We know that ion i can move from the solution (2) into the membrane (3); thus, according to Eq. (3), in equilibrium its electrochemical potentials in the two phases must be equal:

$$\bar{\mu}_3^i = \bar{\mu}_2^i = \mu_2^i + z^i F \phi_2 \tag{9}$$

where ϕ_2 is the inner potential in the solution. Similarly, Fermi levels in the semiconductor (5) and the metal (6) are equal. Because we defined metal (6) to be the same as metal (1) (the reference electrode) we can write

$$\bar{\mu}_5^e = \bar{\mu}_1^e = \mu_1^e - F \phi_1 \tag{10}$$

There is an equilibrium in the metal between the cations and electrons

$$M_1 \rightleftarrows M_1^+ + e_1$$

for which we can formally write

$$\mu_1^M = \mu_1^{M+} + \mu_1^e \tag{11}$$

Substituting for μ_1^e in Eq. (10) we have

$$\bar{\mu}_5^e = \mu_1^M - \mu_1^{M+} - F \phi_1 \tag{12}$$

Combining Eqs. (8), (9), and (12) and rearranging yields

$$\phi_5 - \phi_3 = \frac{1}{F}(\mu_5^e - \mu_1^M + \mu_1^{M+}) - \frac{1}{z^iF}(\mu_2^i - \mu_3^i) + (\phi_1 - \phi_2) \quad (13)$$

The first term on the right-hand side of Eq. (13) represents the contact potential between semiconductor and metal, which can be written

$$\phi_5 - \phi_1 = \frac{1}{F}(\mu_5^e - \mu_1^e) = \Delta\phi_{cont} \quad (14)$$

The second term can be related to the solution activity of the ion (Nernst equation):

$$\frac{1}{z^iF}(\mu_2^i - \mu_3^i) = E_0^i + \frac{RT}{z^iF} \ln a_2^i \quad (15)$$

where a_3^i (activity of species i in phase 3) is assumed to be constant and is included in term E_0. Finally, $\phi_1 - \phi_2$ is the reference electrode potential E_{ref}. Equation (13) can now be written as

$$\Delta\phi_{3/5} = \phi_3 - \phi_5 = \Delta\phi_{cont} + E_0^i + \frac{RT}{z^iF} \ln a_2^i - E_{ref} \quad (16)$$

The voltage across the gate insulator $\Delta\phi_{3/5}$ can be superimposed on the externally applied voltage V_G, which has the same meaning and function as defined in the theory of operation of the IGFET (Chapter 1).

$$I_{DS} = \frac{\mu_n C_0 W V_D}{L} \left(V_G + \Delta\phi_{cont} + E_0^i + \frac{RT}{z^iF} \ln a_2^i - E_{ref} \right.$$

$$\left. + \frac{Q_{ss}}{C_0} - 2\phi_F + \frac{Q_B}{C_0} - \frac{V_D}{2} \right) \quad (17)$$

V_T^* for the ISFET can be defined thus:

$$V_T^* = -\Delta\phi_{cont} - E_0^i - \frac{Q_{ss}}{C_0} + 2\phi_F - \frac{Q_B}{C_0} \quad (18)$$

The inclusion of the term E_0^i (but not E_{ref}) in the threshold voltage V_T^* is rather arbitrary. The reason it is done here is that the membrane is physically part of the ISFET. On the other hand, the reference electrode is a completely separate structure, which is necessary for, but unrelated to, the operation of an ISFET. The final equation for the drain current of the ISFET sensitive to the activity of ions i is then

$$I_D = \frac{\mu_n C_0 W V_D}{L} \left(V_G - V_T^* \pm \frac{RT}{z^iF} \ln a_2^i - E_{ref} - \frac{V_D}{2} \right) \quad (19)$$

for operation in the nonsaturation region and

$$I_D = \frac{\mu_n C_0 W}{2L} \left(V_G - V_T^* \pm \frac{RT}{z^i F} \ln a_2^i - E_{ref} \right)^2 \qquad (20)$$

for operation in the saturation mode.

A. NEUTRAL SPECIES

In light of the proven success of the hydrogen-sensitive field effect transistor (see Chapter 1) it is tempting to speculate that by inserting a chemically reactive layer anywhere in the CHEMFET structure one would be able to obtain a new sensor for that reacting species. This possibility requires further analysis. Let us consider the gate structure shown in Fig. 8 and designate phase n_2 as the chemically reacting layer. Assume that interfaces A and B are resistive and that the communicating species is an electron. The Fermi level in each phase n_1 through n_3 is

$$\bar{\mu}^e = \mu^e - F\phi \qquad (21)$$

The potential difference between n_1 and SC, V_{INS}, which determines the magnitude of the drain current, is then

$$V_{INS} = \phi_{SC} - \phi_{n_1} = \phi_{n_2} - \phi_{n_1} + \phi_{n_3} - \phi_{n_2} + \phi_{SC} - \phi_{n_3} \qquad (22)$$

However, because both the n_1–n_2 and n_2–n_3 interfaces are resistive (i.e., Fermi levels in phases n_1, n_2, n_3 are equal), ϕ_{n_2} cancels out in Eq. (22). Therefore, even if the electron affinity changes in n_2 as the result of the gas interaction, there is no effect on $\phi_{SC} - \phi_{n_1}$ and the change of ϕ_{n_2} cannot be measured. In other words, a change of potential difference at

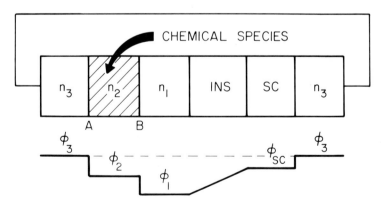

Fig. 8. Potential distribution in the CHEMFET gate in which layer n_2 reversibly interacts with neutral species.

interface A is compensated by an equal but opposite change at interface B.

Let us now consider the case that one interface, for instance, A, is capacitive. In that case the potential difference between phase n_2 and the semiconductor is distributed across two capacitors, C_A and C_{INS}, with phase n_1 being electrically equivalent to an equipotential connection between the two capacitors. The potential drop across the interface A will usually be small compared to the potential drop across the insulator because $C_A \gg C_{INS}$.

The potential distribution between the capacitors created by interface A and the insulator is then

$$C_A V_A = C_{INS} V_{INS} \tag{23}$$

or

$$\Delta V_{INS} = \Delta \phi \frac{C_A}{C_{INS}} \tag{24}$$

Thus the change in the inner potential of phase n_2 will affect the overall potential distribution, which will then translate into the change of the drain current through the drain current system.

The same argument can be made for two ionic species communicating across the interfaces B and A. It can be generalized that a transistor can be made to respond to a chemical stimulus by a neutral species provided that one of the interfaces of the reactive layer is capacitive or, if it is resistive, that it maintains a constant potential. It can be readily seen that the second requirement is also satisfied by an ISFET, in which case the change of activity of one ion affects the solution–membrane interface, but not the reference electrode–solution interface.

B. CHEMFETs with Nonpolarized Interfaces—ISFETs

The ion selectivity of ISFETs is obtained by placing a membrane material that is selectively permeable to the ion of interest over the gate insulator; in Fig. 7, this membrane is the phase 3. In principle, it can be any of the ion-selective membranes used in conventional electrodes. The practical limitation is that a suitable method of deposition of this material must exist. This last condition has proved to be troublesome, for example, in preparation of a fluoride-sensitive ISFET, and so far only hybrid devices sensitive to F⁻ have been described (Fjeldy and Nagy, 1980). There also has been a tendency to deposit various ion-selective materials by techniques compatible with standard integrated circuit fabrication technology, for example, by sputtering (Topich *et al.*, 1978; Esashi and Matsuo, 1978), by chemical vapor deposition, or by combining some of the electroactive components of conventional ISE membranes into photoresist (Wen *et al.*, 1980). These approaches would certainly facilitate the fabrication of ISFETs, but the electrochemical properties of these membranes are often compromised, so that the overall performance is degraded. While the development of new techniques of material application that are compatible with integrated circuit fabrication processes is highly desirable, these techniques must be such that the ion-selective materials will not be degraded during the deposition or the encapsulation procedures.

1. ISFETs with Thin-Film Ion-Selective Membranes

The first ISFETs reported in the literature were pH-sensitive devices with SiO_2 (Bergveld, 1970) or $Si_3 N_4$ (Matsuo and Wise, 1974) serving as both the gate insulator and the ion-selective membrane. Transistors with thick polymeric ion-selective membranes were reported later (Moss *et al.*, 1975). A pH-sensitive structure has been studied by several authors, either as a transistor or as an electrolyte–insulator–semiconductor capacitor, by using capacitance–voltage measurement techniques. Unfortunately, some of these studies were done over a broad pH range; instead of using pH buffers, the response curve was obtained by titrating a strong acid with a strong base. Under those conditions, the pH range 5–9 has extremely low buffer capacity and the actual pH values are unreliable. Thus, despite these efforts, the origin of the pH response is not clear. It suffices to say here that a pH-dependent charge is generated by ionizable groups at the surface of the insulator or within the hydrated surface layer of an otherwise insulating material. This charge gives rise to the pH-related signal of the transistor. We must realize, however, that it would not be possible to make a glass electrode sensor out of Si_3N_4 or SiO_2 because in bulk these materials are insulating. However, these materials work well as thin, pH-sensitive layers in the ISFET configuration. Other thin oxide films have been used in pH ISFETs: Ta_2O_5, Al_2O_3, TiO_2. The common denominator of all these devices is the fact that the pH-dependent charge is coupled directly to the channel charge in the transistor. Thin layers of aluminosilicates or borosilicates have been deposited on top of the silicon nitride layer by chemical vapor deposition (Esashi and Matsuo, 1978). These devices exhibit a potential change of 55 mV/pNa in the range of pNa 0–3; this change is somewhat lower than that of a good sodium-selective electrode. Similarly, an Na^+-sensitive ISFET was prepared by Li^+ and Al^{3+} ion implantation into a plasma-deposited SiN layer

(Sanada *et al.*, 1982). The response of these devices to Na^+ ions (30 mV/ pNa, range pNa 0–3) was again degraded compared to that of a conventional ISE or Na^+ ISFET.

2. ISFETs with Thick Membranes

The most successful technique for fabrication of different ISFETs so far has proved to be solvent casting of the membrane. Both homogeneous (McBride *et al.*, 1978; Oesch *et al.*, 1981) and heterogeneous (Shiramizu *et al.*, 1979) membranes have been applied in this way. Transistors sensitive to hydrogen ions (Janata and Huber, 1980), potassium and calcium ions (McBride *et al.*, 1978), sodium and ammonium ions (Oesch *et al.*, 1981), and phenobarbital anion (Covington *et al.*, 1982) have been made by solvent casting. It is obvious that ISFETs with other polymeric membranes can be made in a similar way.

There has been, however, one major problem with these devices. The adhesion of the membrane to the silicon nitride surface of the chip and to the surrounding encapsulant is rather poor; even with membranes 100 μm thick a current leakage path develops between the membrane and the encapsulant. It also has been shown that the surface of wet silicon nitride in these devices is conducting (Cohen and Janata, 1983a); the combination of these two effects leads to electrical shunting of the membrane potential. This problem is substantially aggravated when the device is subjected to some mechanical stress, such as may occur during *in vivo* measurements. In that particular situation McKinley *et al.* (1981) anchored the membrane to the epoxy by a small ring of polyvinyl chloride (PVC) with a low plasticizer content. Another solution was to cast the ion-selective membrane over an area much larger than the gate, preferably over the entire end of the encapsulated probe. Although devices made in this way have been shown to work for over 2 mo in solution (McBride *et al.*, 1978) this approach is totally unsuitable for multisensor fabrication. Recently, Blackburn and Janata (1982) used a new approach to this problem, in which a three-dimensional structure is built above the transistor gate (Fig. 9). The entire area of the chip is covered with polyimide, except the gate itself and the bonding pads. The suspended polyimide mesh provides a mechanical support for the membrane. Also, PVC-based membranes have better adhesion to polyimide. The superior performance of suspended mesh ISFETs is shown in Fig. 10. The most important aspect of this improvement is that the area of the membrane is the same as the area of the mesh and, therefore, multisensor chips with satisfactory long-term performance can be prepared. The details of this fabrication procedure will be discussed in Chapter 3.

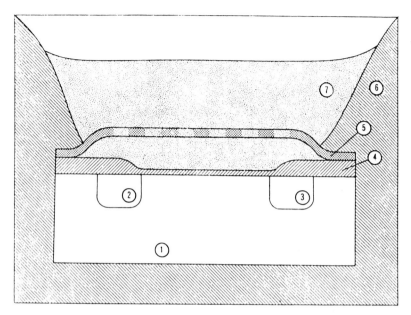

Fig. 9. Schematic diagram of suspended mesh ISFET. (1) Substrate; (2 and 3) drain and source; (4) insulator; (5) suspended polyimide mesh; (6) encapsulant; (7) ion-selective membrane. (From Blackburn and Janata, 1982. Reprinted by permission of the publisher, The Electrochemical Society, Inc.)

C. CHEMFETs with Polarized Interfaces

First, we have to consider a semantic point: it has been suggested (Lauks, 1981) that all CHEMFETs are polarized devices because they contain a capacitor in the gate structure. We have also said (Section I, B) that no matter how many layers there are in the gate structure the interface between the dry insulator and the adjacent layer is always polarized. While this is true it does not help us understand the interaction of the sample (solution) with the CHEMFET. From that point of view it is important to understand the nature of the solution–device interface. If this interface is polarized then the transistor can be used to measure excess charge at this interface. If this interface is nonpolarized it can be used to measure the difference between the inner potential of the solution and the inner potential of the membrane. In that case, the relationship between the activities of the participating ions and the potential is described by some form of the Nernst equation, Eq. (15).

The polarizability of an interface depends on the type of solid phase, the composition of the solution, and the interfacial potential. In the ensuing discussion we shall consider first the case of an ideally polarized

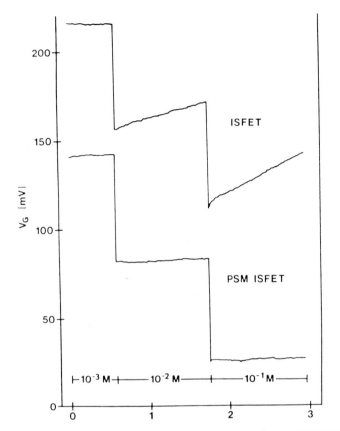

Fig. 10. Typical potassium ion standardization curves for the normal ISFET and suspended mesh (SM) ISFET after 7 days in solution. (From Blackburn and Janata, 1982. Reprinted by permission of the publisher, The Electrochemical Society, Inc.)

interface, that is, one at which there is no charge transfer. This situation is described by the Gibbs–Lippmann equation, which at constant pressure and temperature is

$$-d\gamma = q^m \, dE + \sum_i \Gamma_i \, d\mu_i \tag{25}$$

where γ is surface energy, q^m is charge on the solid phase (here metal*), E is the interfacial potential, and Γ_i is the relative surface excess, defined as

* In this discussion we are considering a CHEMFET with a thin layer of metal (e.g., gold) that is electrically floating.

$$\Gamma_i = \Gamma_i^* - \frac{n_i}{n_0} \Gamma_0^* \tag{26}$$

Thus, the relative surface excess Γ_i is defined in terms of the surface concentrations Γ_i^* and Γ_0^* of species i and a reference species 0, usually the solvent. Compared to the Nernst equation, the Gibbs–Lippmann equation has an additional degree of freedom (Mohilner, 1966). Consequently, the relationship between the interfacial charge and the bulk activity of adsorbing species must be studied either at constant charge or at constant interfacial potential. The important experimental difference between a CHEMFET with a polarized interface and a conventional nonpolarized (ion-selective) electrode is that the CHEMFET has an additional experimental parameter that can be controlled, the drain current. In other words, this CHEMFET can be operated either at constant gate charge (constant drain current) or at a constant applied potential (variable drain current). The drain current, therefore, can be looked at as a means for probing the charge density at the semiconductor plate of the solution–insulator–semiconductor capacitor.

1. Measurement of Potential at Constant Interfacial Charge

The amount of excess charge at a polarized electrode–solution interface is normally determined by measuring the differential capacitance over a range of potentials and then integrating twice with respect to the potential, using the value of differential capacitance at a sufficiently negative potential as the integrating constant (Mohilner, 1966). If the relationship between the bulk activity of the adsorbing species and the excess interfacial charge, the so-called Esin–Markov relationship, is being studied, the above procedure must be repeated for every solution composition. Because of the laboriousness of this procedure, the practical analytical value of this relationship has never been exploited.

The Esin–Markov coefficient

$$\left(\frac{\partial E}{\partial \ln a}\right)_{q^{\mathrm{m}}} = -RT \left(\frac{\partial \Gamma_i}{\partial q^{\mathrm{m}}}\right)_{a_i} \tag{27}$$

can be derived (Joshi and Parsons, 1961) directly from the Gibbs–Lippmann equation. The right-hand side of Eq. (27) depends on the specific form of the adsorption isotherm. Notice that Eq. (27) requires that the charge on the metal be kept constant. This condition can be met experimentally by operating the CHEMFET at constant drain current, as will be shown below.

In the case of a strong specific adsorption the drain current can be expressed as

$$I_D = \frac{W\mu_n C_0 V_D}{L} \left(V_G - V_T^* - E_{ref} + \frac{q_i}{C_0} - \frac{V_D}{2} \right) \tag{28}$$

or, for a device operated in saturation,

$$I_D = \frac{W\mu_n C_0}{2L} \left(V_G - V_T^* - E_{ref} + \frac{q_i}{C_0} \right)^2 \tag{29}$$

(See Chapter 1 for definitions of m and n.) In this case constancy of the interfacial charge is not required; however, it is still preferable to operate the device in the constant-current mode because the change of the interfacial potential can be obtained directly.

The charge and potential profiles across the CHEMFET gate with a polarized interface are shown in Fig. 11. The applied gate voltage V_G is the sum of the reference electrode potential ϕ_R, double-layer potential ϕ_{dl}, potential drop across insulator ϕ_i, and surface potential ϕ_s. Thus

$$V_G = \phi_R + \phi_{dl} + \phi_i + \phi_s \tag{30}$$

The double-layer potential ϕ_{dl} is identical to the potential of the polarized electrode E in Eq. (27).

$$\phi_{dl} = E = V_G - \phi_i - \phi_R - \phi_s \tag{31}$$

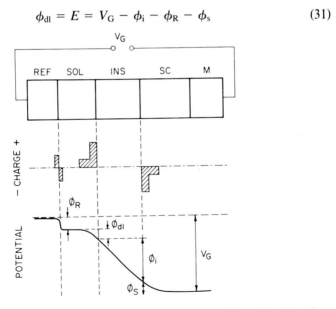

Fig. 11. Charge and potential profiles across CHEMFET gate with polarized interface. REF, reference electrode; SOL, solution; INS, insulator; SC, semiconductor; M, metal. (Reprinted with permission from Janata and Huber, 1980.)

The potential ϕ_i is related to the gate capacitance and the total charge q_i at the semiconductor side of the insulator:

$$\phi_i = \frac{q_i d}{\varepsilon_i} \tag{32}$$

where d is the insulator thickness and ε_i is its permittivity. It is important to note that when charge q_i is held constant (therefore ϕ_i is constant), the change of ϕ_{dl} with bulk activity of the adsorbing species is the Esin–Markov coefficient, because ϕ_R and ϕ_s are constant.

The principle of this direct measurement of interfacial excess charge is shown in Fig. 12. For a given composition of the solution and a given applied gate voltage V_G, the value of the drain current is set at $I_D = \text{const}$ (curve 1). When more cations adsorb at the interface the potential ϕ_{dl} increases, which results in a corresponding increase of the drain current (curve 2). In order to maintain a constant interfacial charge the drain current is held constant at its original value by adjusting V_G (curve 3). It can be seen that the drain current thus serves as a third experimental variable, the first two being bulk activity and applied gate voltage. Therefore, the three degrees of freedom in the Gibbs–Lippmann equation, charge, potential, and activity (at constant P and T), are matched by three

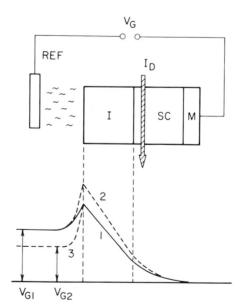

Fig. 12. Measurement of interfacial charge with CHEMFET. (Reprinted with permission from Janata and Huber, 1980.)

experimental variables. In this way the transistor is used in a constant-current mode.

In order to verify this theory the adsorption of iodide from a 0.1 M NaF solution onto a thin layer of gold deposited on the top of the gate insulator has been studied (Cohen and Janata, 1983b). Because this gold layer is electrically floating and uncharged, the potential profile is essentially the same as that shown in Fig. 11, except for an interposed layer of gold, which is equipotential. The Esin–Markov relationship discussed above holds for this structure. However, two kinds of experimental difficulties have been encountered with the thin gold layer structure, both related to the materials used in these devices. First, titanium, which is used as a bonding underlayer beneath the gold, diffuses along the grain boundaries to the top of the gold. This results in the formation of a redox couple at the interface, which then "pins" its potential. The addition of iodide produces a step change of potential, which then returns to the former potential of the gold surface contaminated with titanium (the pinning potential). The time constant of this process depends on the amount of titanium in the surface, which is time-variable and difficult to control.

The second problem encountered was of a more general nature. It was found (Cohen and Janata, 1983a) that hydrated silicon nitride exhibits relatively high surface conductivity (1.6×10^{-15} ohm^{-1} square). This conductivity was sufficiently high to dissipate the charge from the gold layer to the surrounding "insulator." Again, the time constant was dependent on the processing parameters and the history of each device and on the degree of hydration of the surface. However, it must be stressed that silicon nitride is a perfect insulator in the perpendicular direction; the described difficulties related only to its lateral surface conductivity.

Both problems have been circumvented by attaching a short piece of gold wire to the floating metal gate and encapsulating the whole structure with a thick layer of a high-grade epoxy (Cohen and Janata, 1983b) (Fig. 13). After mechanical and electrochemical cleaning of this hybrid electrode, stepwise addition of iodide produced the change of interfacial potential (Fig. 14). The plot of potential change against activity of iodide was linear with a slope of 53.6 mV/log C. The step changes of potential in Fig. 14 nevertheless show a small drift that was attributed to the presence of adsorbed oxygen, which reduces the charge transfer resistance of the interface and causes potential pinning.

2. Measurement of Charge at Constant Potential

By direct cross-partial differentiation of the Gibbs–Lippmann equation, Eq. (25), we can observe the dependence of the metal charge on the

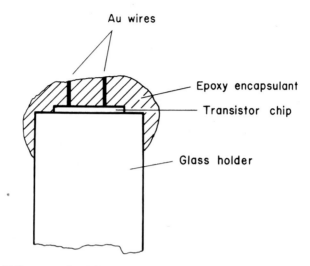

Fig. 13. Fully encapsulated CHEMFET with two gold wires. (Reprinted with permission from Cohen and Janata, 1983b.)

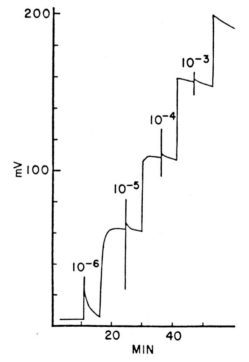

Fig. 14. Response of gold wire–gate transistors to successive additions of NaI. Supporting electrolyte was 0.1 M NaF. (Reprinted with permission from Cohen and Janata, 1983b.)

bulk activity of the adsorbing ion at constant interfacial potential

$$\left(\frac{\partial q^{m}}{\partial \ln a_{i}}\right)_{E} = RT \left(\frac{\partial \Gamma_{i}}{\partial E}\right)_{u_{i}} \tag{33}$$

Since this relationship is derived from the same equation as the Esin–Markov coefficient it cannot contain any new information. However, it is directly accessible experimentally by using a CHEMFET with a polarized interface. Furthermore, it is directly related to the Esin–Markov coefficient by the thermodynamic relationship that holds for any three variables of state x, y, z (Sears, 1953)

$$\left(\frac{\partial x}{\partial y}\right)_{z}\left(\frac{\partial y}{\partial z}\right)_{x}\left(\frac{\partial z}{\partial x}\right)_{y} = -1 \tag{34}$$

Thus, for charge, potential, and activity we obtain from Eq. (25)

$$\left(\frac{\partial E}{\partial \mu_{i}}\right)_{q^{m},j}\left(\frac{\partial \mu_{i}}{\partial q^{m}}\right)_{E}\left(\frac{\partial q^{m}}{\partial E}\right)_{\mu_{i}} = -1 \tag{35}$$

If q^{m} is a smooth, continuous, and nonsingular function of μ_{i} we can measure its inverse. The third term in Eq. (35) is the differential capacitance C_{d}; therefore

$$\left(\frac{\partial E}{\partial \ln a_{i}}\right)_{q^{m},j} = -\frac{1}{C_{d}}\left(\frac{\partial q^{m}}{\partial \ln a_{i}}\right)_{E} \tag{36}$$

The circuit shown in Fig. 15 (Cohen and Janata, 1983c) is used to measure the variation of charge with bulk activity of iodide at a constant interfacial potential.

Two methods of measurement of the interfacial charge have been demonstrated, one with constant potential and one with constant charge. In both cases the model system chosen for experimental verification was the specific adsorption of iodide on gold. It can be concluded that the steady-state determination of charge in either mode will be difficult because of the nonideal behavior of the polarized interface; the finite charge transfer resistance in parallel with the differential capacitance gives rise to a time constant, which determines the rate of leakage of the charge from the floating gate metal either to its surrounding or to the solution. The strategy of the development of analytical probes based on direct measurement of the interfacial charge with transistors should, therefore, be based on some form of transient concentration step measurements in which the time constant of this concentration step would be short compared to the time constant of the leakage of the charge from the interface. It is also clear from our studies that the choice of materials from which the device is fabricated is very important in this respect.

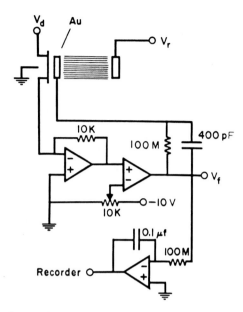

Fig. 15. Circuit diagram for measurement of adsorbed charge at constant applied potential. (Reprinted with permission from Cohen and Janata, 1983c.)

D. CHEMFETs WITH NONCONVENTIONAL MEMBRANES

In the previous two sections we were dealing with two extreme electrochemical cases: CHEMFETs with totally nonpolarized membrane–solution interfaces and transistors with ideally polarized membrane–solution interfaces. The main practical distinction between the two is that the latter can be used to measure excess interfacial charge.

High input resistance and a fixed input capacitance make the CHEMFET an ideal amplifier for membranes that fall between these two extremes. While this possibility increases the number of potential new sensors that could be developed, it also presents a danger; it is often difficult to unravel the mechanism by which the electrical signal is generated in response to a chemical stimulus. The specific response is usually obtained only if the experimental conditions are rigorously controlled (Solsky and Rechnitz, 1981). An example of such a situation is the membrane that is antigenic to the syphilis antibody (Aizawa *et al.*, 1977), and which was used in a potentiometric "immunosensor." This membrane was cast on a CHEMFET (Collins and Janata, 1982) gate and the response to the venereal disease (VDRL+) serum was obtained (Fig. 16). Unfortunately, the same membrane responded equally well to changes in concentration of sodium chloride (Fig. 17) or any small inorganic ion, including

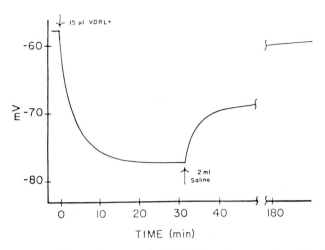

Fig. 16. Response of a transistor coated with VDRL–antigen membrane to addition and dilution of VDRL+ serum. Temperature, 37.0°C; 2.0 ml of 0.15 M NaCl. (Reprinted with permission from Collins and Janata, 1982.)

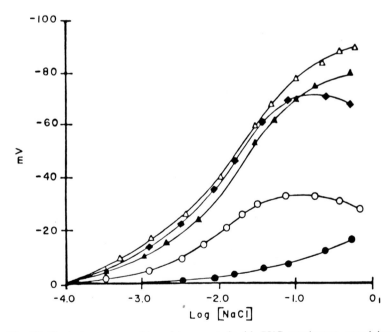

Fig. 17. Typical response of transistors coated with PVC membranes containing the following antigenic components: (●) 0.1% cardiolipin and 2% cholesterol; (○) 0.1% cardiolipin and 0.4% lecithin; (▲) 0.4% lecithin and 2.0% cholesterol; (◆) 2.0% cholesterol; (▽) total antigen. The initial solution was 150 mM HEPES. (Reprinted with permission from Collins and Janata, 1982.)

hydrogen and hydroxyl ions. Furthermore, any surface-active molecules produced a response as well. In order to explain this behavior, a thin (1–10 μm) membrane was clamped between two saline solutions and a current–voltage curve was measured (Collins and Janata, 1982). It was found that ionic d.c. current can pass through this membrane and, therefore, that the solution–membrane interface cannot be polarized. Since the equilibrium potential difference at a nonpolarized interface depends only on the difference between the inner potentials of the solid and solution phase and not on the adsorption (Buck, 1982), the origin of these changes had to be kinetic.

It is important to realize the difference between these membranes and a regular, well-behaved ion-selective membrane in terms of exchange current density. A good ion-selective membrane has an exchange density in the range of 10^{-2}–10^{-6} A/cm², which is due mainly to one ion (Camman, 1978; Koryta et al., 1977). On the other hand, the exchange current density for a mercury electrode in, for example, NaF solution is 10^{-12} A/cm². Such an electrode is normally regarded as polarized. If the overall exchange current density is low ($10^{-6} > i_0 > 10^{-10}$ A/cm²), and includes current contributions from many ions, a mixed potential is established at the membrane–solution interface. This situation for only two ionic species is shown in Fig. 18. The ion current for each ionic species that crosses the membrane–solution interface can be expressed as

$$i_j^s = z_j F k_j a_j^s \exp\left[\frac{-\alpha_j z_j F(E_m - E_j)}{RT}\right] \tag{37}$$

for ions j entering from solution, and

$$i_j^m = z_j F k_j a_j^m \exp\left[\frac{(1 - \alpha_j)z_j F(E_m - E_j)}{RT}\right] \tag{38}$$

for ions leaving the membrane, where z_j is the ionic valence, k_j is the heterogeneous rate constant, and a_j^s and a_j^m are the surface concentrations of ion j on the solution and membrane side, respectively. Terms F, R, and T have their usual meaning and α_j is the symmetry coefficient. The equilibrium potential for species j is E_j and the mixed potential E_m corresponds to the condition of zero net current

$$\sum_j i_j^s = \sum_j i_j^m \tag{39}$$

If any of the partial currents is affected by the adsorption of proteins or other molecules the E_m will shift. This effect can work through the change of α, k, or surface concentration (Buck, 1982; Collins and Janata, 1982; Koryta et al., 1977; Blackburn and Janata, 1984). Because of the presence

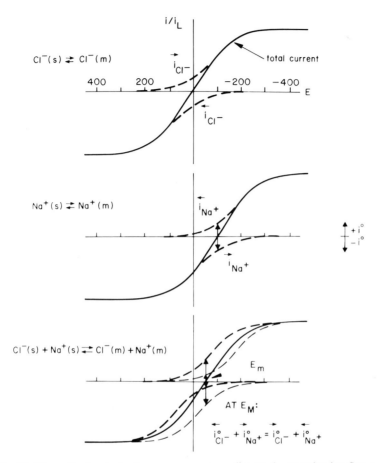

Fig. 18. Normalized polarization curves corresponding to the negative ion flux out of ($\rightarrow$) and into ($\leftarrow$) the membrane. Symbols (s) and (m) designate solution and membrane, respectively; E_M is the mixed potential.

of parasitic leakage currents, these membranes would be hopelessly unstable when used in coated wire or conventional ISE configurations. However, with CHEMFETs they give a deceptively stable reading (Collins and Janata, 1982). Unfortunately, all the ions that contribute to the overall exchange current density will interfere with the potential of this probe. The same mechanism can be invoked to explain the pH dependence of a PVC membrane whose surface is impregnated with carboxylic groups. Again, this membrane had a mixed potential response to small ions, a response that was modulated by the protonation equilibrium at the surface (Janata and Huber, 1980). The same mechanism was apparently also involved in the previously reported "immunoelectrode" sensitive to various glucopyranosides (Janata, 1975).

There is, however, a report (Fujihira *et al.*, 1980) of a transistor with a thin (1000 Å) parylene film that may have formed a substantially polarized membrane–solution interface. If this is so, then such a device could serve as the basis of a true immunochemically sensitive transistor.

E. TEMPERATURE DEPENDENCE OF THE ISFET

The strong similarity between the IGFET and the ISFET can be exploited in the analysis of the temperature dependence of the ISFET.

Because both drain current Eqs. (19) and (20) depend only on the quantities $\beta = \mu_n C_0 (W/L)$, V_T, and applied voltages, any temperature dependence must be through the quantities β and V_T. Since C_0 is the gate insulator capacitance per unit area,

$$C_0 = \frac{\varepsilon_i}{d_{ins}} \tag{40}$$

where ε_i is the permittivity of the insulator and d_{ins} is the thickness of the insulator. These parameters are relatively constant with temperature in the range of interest, so any temperature variation of β must be through μ_n, the charge carrier mobility in the surface inversion layer. This quantity has been studied as a function of temperature (Vadasz and Grove, 1966). In the temperature region of practical interest, -55 to $+120°C$, the variation of μ_n with temperature is satisfactorily approximated by a T^{-1} dependence. Therefore, the variation of mobility with absolute temperature T will introduce a T^{-1} dependence into I_D.

Of the terms in the equation for V_T, the temperature dependence of ϕ_f and ϕ_B must be considered. The fixed oxide charge Q_{ss} is observed to be relatively independent of temperature over the temperature range of interest. When dealing with the IGFET, ϕ_{ms} can be considered independent of temperature. However, when the expression for V_T is modified for the ISFET, the terms that replace ϕ_{ms} in the threshold equation are temperature-dependent [see Eq. (18), Section II].

The ISFET drain current [Eq. (19)] may now be differentiated with respect to temperature to show the temperature dependence:

$$\frac{dI_D}{dT} = I_D \left\{ \frac{1}{\mu_n} \frac{d\mu_n}{dT} - \frac{V_D[dV_T/dT - d(EMF)/dT]}{V_G - V_T^* \pm (RT/z^i F)(\ln a^i) - E_{ref} - V_D/2} \right\} \tag{41}$$

The term $d(EMF)/dT$ is the temperature dependence of the electrochemical cell consisting of the reference electrode, the electrolyte, and the ion-sensitive membrane. The term dV_T/dT accounts for the temperature dependence of the semiconductor band gap and bulk Fermi level. It is the same as the temperature variation of the IGFET threshold equation; dV_T/dT is always negative for n-channel structures.

Inspection of Eq. (41) shows that, given $d\mu_n/dT$ always negative and the term $[dV_T/dT - d(\text{EMF})/dt]$ always negative (Janata and Moss, 1976), it is possible to choose an operating point around which I_D is, in the first order, independent of temperature. However, it is known that the term $d(\text{EMF})/dT$ varies with the activity of the measured ion and is equal to zero only for one value of activity, the so-called isopotential. The effect of temperature on the drain current can therefore be minimized by the proper selection of V_G, but can never be eliminated completely.

III. Dynamic Characteristics

A. TRANSIENTS DUE TO CHANGE OF GATE VOLTAGE

It has been postulated (Bergveld, 1972; Janata and Moss, 1976) that the in situ impedance transformation will shorten the response time of ISFETs as compared to the equivalent ion-selective electrodes. In ISEs the time constant RC is independent of the membrane geometry, although the individual values of the bulk resistance R_B and bulk capacitance C_B vary with the thickness of the membrane. In ISFETs we have to consider these parameters in relation to the input capacitance of the transistor gate.

Let us consider the equivalent circuit shown in Fig. 19a, in which C_B and R_B represent the bulk membrane capacitance and resistance, respectively. The amplifier input voltage V_{GS} appears across the input capacitor C_{GS}. The response function of this circuit to a step input function is (Haemmerli et al., 1982a)

$$V_{GS}(t) = V_i \left(1 - \frac{C_{GS}}{C_B + C_{GS}} e^{-t/R_B(C_B + C_{GS})}\right) \quad (42)$$

For $t = 0$

$$V_{GS}(0) = V_i \frac{C_B}{C_B + C_{GS}} \quad (43)$$

In other words, the bulk membrane capacitance and the input capacitance form a capacitive divider in which the output voltage $V_{GS}(0)$ appears instantaneously when the input voltage is applied. If $C_{GS} \ll C_B$ then $V_{GS}(0) = V_i$. On the other hand, $\lim_{C_{GS} \to \infty}[V_{GS}(0)] = 0$. This is usually the case with conventional ISEs, where the lead capacitance is large and must be added to the input capacitance for the purpose of the analysis. However, with the ISFET the situation is markedly different because the input capacitance is comparable to the bulk membrane capacitance. Therefore, a response curve with two time constants is obtained with the ISFET, as

a

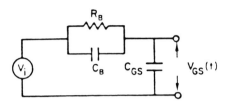

b

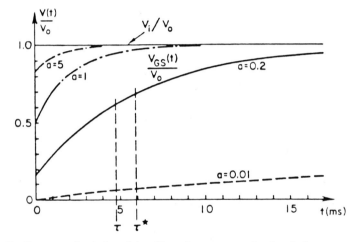

Fig. 19. Computer simulation of the effect of geometry on the electrical response time of an ISFET. (a) Equivalent circuit; (b) simulated response. Parameter $a = C_B/C_{GS}$. (From Haemmerli *et al.*, 1982a. Reprinted by permission of the publisher, The Electrochemical Society, Inc.)

shown in Fig. 19b. This behavior is predicted by the theoretical model represented by the equivalent circuit shown in Fig. 19a. This model has been verified experimentally by measuring the response of a Cl^- micro ISFET (MISFET) to an electrical step of the gate voltage (Fig. 20). The equivalent circuit values of $R_B = 3.2 \times 10^{10}$ ohms and $C_B = 0.8$ pF yield the time constant of 2.5 ms. Time constants in the range of milliseconds were found (Smith *et al.*, 1980) for other ISFETs with polymeric membranes. Time responses on the order of tens and hundreds of milliseconds were found for ion-selective micro ISFETs (Haemmerli *et al.*, 1982b), while a pH ISFET with a bare silicon nitride gate has a response time of microseconds. However, it must be remembered that the response to a

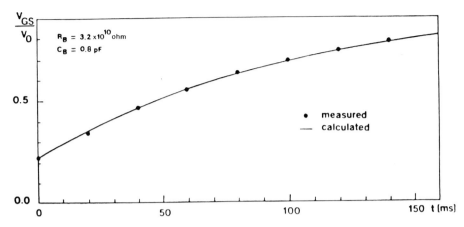

Fig. 20. Electrical response time [measured (●) and calculated (solid line)] of Cl^- micro ISFET. (Reprinted with permission from Haemmerli *et al.*, 1982b.)

concentration step is usually dominated by the mass transport through the unstirred solution boundary layer. In that respect the time response of the whole system with an ISFET should not be significantly different from that of the equivalent ISE.

A special technique based on flow injection analysis (FIA) has been designed for testing micro ISFETs (Haemmerli *et al.*, 1982c). A sharp concentration transient created in the FIA system proved to be ideal for testing the dynamic characteristics of these devices, which have a fragile submicrometer-diameter electrode tip. The signal transients corresponding to concentration increase and concentration decrease are shown in Fig. 21. Typical values obtained by this technique were 240 ms for the K^+ MISFET and 490 ms for the Cl^- ISFET.

B. TRANSIENTS DUE TO CHANGE OF DRAIN CURRENT

While the step change in gate voltage produces an exponentially changing drain current, there is a pronounced overshoot (undershoot) when the drain voltage is increased (decreased) (Smith *et al.*, 1980) (Fig. 22). This behavior would be encountered, for example, during multiplexing of a multi-ISFET chip when the individual devices would be periodically switched on and off, or during an a.c. drain voltage operation. It is caused by the finite mobility of charge in the membrane and can be rationalized in the following way. For a given value of drain current, there is a distribution of charge as shown in Fig. 23. In the segment dy there is the same amount of charge the semiconductor channel and in the ion-selec-

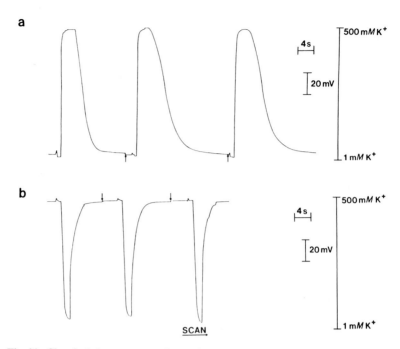

Fig. 21. Chemical time response of K⁺ micro ISFET to (a) step up and (b) step down of the concentration of K⁺ ion. The background electrolyte was 0.1 M NaCl. (Reprinted with permission from Haemmerli *et al.*, 1982c.)

tive membrane. When the channel current, and consequently the channel charge distribution, is changed in a step fashion, the charge on the membrane side of the gate capacitor is changed accordingly. This change produces a lateral transient current in the membrane. Because the charge mobility in ion-selective membranes is relatively low, a step change in drain voltage results in a non-steady-state response (overshoot), which relaxes with a time constant to the steady state. This behavior can also be seen in the drain current-drain voltage relationship (Fig. 24). If V_D is stepped from C to D the effective V_G increases as well, to give a value of I_D at point D''. As the charge relaxes to its steady-state distribution the I_D decreases from D'' to E in a process with two time constants. This situation does not exist in transistors with metal gates; the mobility of electrons in metal is higher than the mobility of charge carriers in the channel and, therefore, any charge redistribution in the channel is matched exactly in the metal. Accordingly, the overshoot can be reduced by interposing a thin layer of gold between the membrane and the insulator.

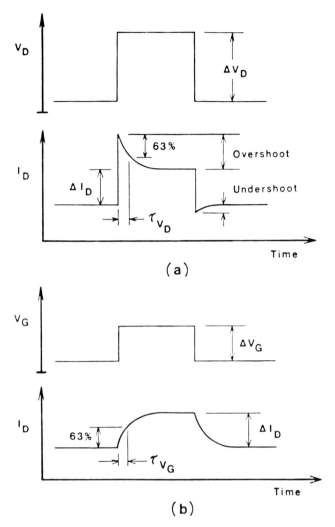

Fig. 22. Transients of the drain current I_D caused by the step of (a) drain-to-source voltage V_D and (b) gate voltage V_G. (From Smith *et al.*, 1980. Reprinted by permission of the publisher, The Electrochemical Society, Inc.)

C. EQUILIBRIUM NOISE IN ISFETs

The small size of the ion-selective membrane and the small and invariable input capacitance make the ISFET a potentially suitable tool for the study of stochastic processes within the device (Haemmerli *et al.*, 1982a).

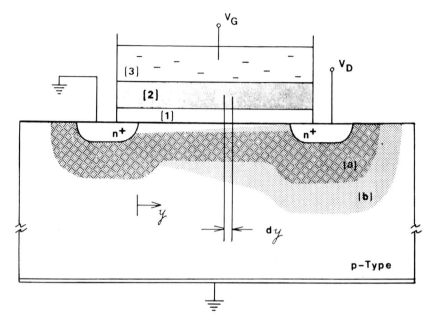

Fig. 23. Charge distribution in the channel of an ISFET for two values of V_D, $V_D(b) > V_D(a)$. (From Smith *et al.*, 1980. Reprinted by permission of the publisher, The Electrochemical Society, Inc.)

The drain current in the channel of an ISFET can be expressed as

$$I_D(t) = \bar{I}_D + i(t). \tag{44}$$

where $\bar{I}_D$ is the average value and $i(t)$ is the random fluctuation (noise) of the drain current. For the purpose of this study, the noise can be formally divided into two parts: the noise originating in the solid state part of the device and the noise associated with the electrochemical components. The former can be studied separately, using IGFETs on the same chip. The schematic diagram for the determination of the spectral density of the current is shown in Fig. 25. Examples of the power spectrum for three ISFETs are shown in Fig. 26. The analysis of the experimental data is done in three steps: first, an equivalent circuit corresponding to the polymeric membrane ISFET is constructed (Fig. 27), in which noise sources V_1, V_2, and V_3 are placed in series with the hypothetical noiseless resistors R_1, R_2, and R_3. Using this model with estimated values for R and C, a power spectrum is digitally stimulated and matched with the experimentally measured spectrum. The lines in Fig. 26, drawn through the measured values, are such simulated theoretical curves. The model RC values

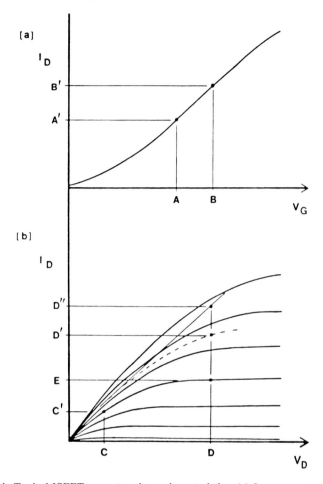

Fig. 24. Typical ISFET current–voltage characteristics. (a) I_D versus gate voltage G_G; (b) I_D versus drain voltage V_D. (From Smith *et al.*, 1980. Reprinted by permission of the publisher, The Electrochemical Society, Inc.)

are then assigned to the membrane bulk and interfacial impedances (Buck, 1978), from which the interfacial exchange current densities and differential capacitances can be estimated. Preliminary results of the measurement of equilibrium noise in ISFETs indicate that this technique yields electrochemical information about ion-selective membranes (e.g., exchange current density) that would be difficult to obtain otherwise. However, the noise contributed by the solid state part of the device must be lower than the noise contributed by the membrane. This proved to be the limiting factor for some membranes used.

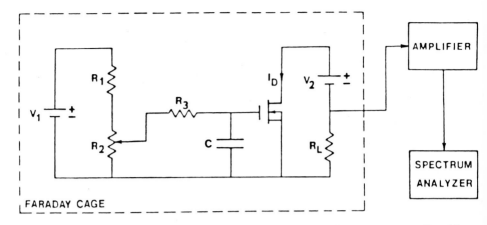

Fig. 25. Experimental arrangement for ISFET and IGFET noise measurements. $R_1 = 25$ kilohms, $R_2 = 0$–50 kilohm potentiometer, $R_3 = 100$ kilohms, $R_L = 10$ kilohms, $C = 1$ μF, $V_1 = 9$ V, and $V_2 = 6$ V. (From Haemmerli *et al.*, 1982a. Reprinted by permission of the publisher, The Electrochemical Society, Inc.)

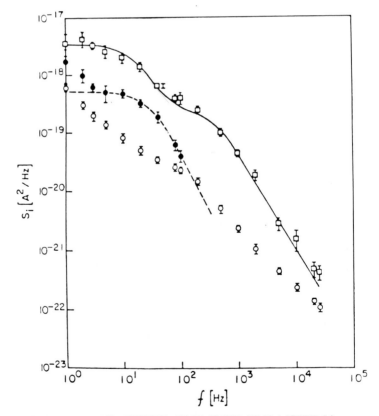

Fig. 26. Noise spectra: (○) pH ISFET; (●) K$^+$ ISFET; (□) Na$^+$ ISFET. Lines represent calculated response. (From Haemmerli *et al.*, 1982a. Reprinted by permission of the publisher, The Electrochemical Society, Inc.)

a
NOISE MODELING

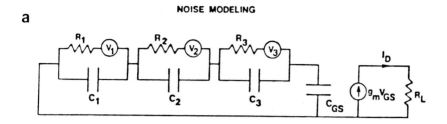

b

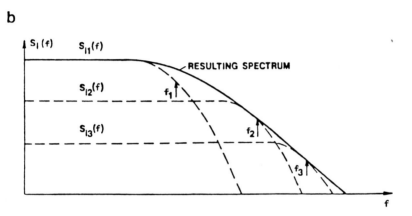

Fig. 27. (a) Electrical equivalent circuit used for modeling. (b) Log–log representation of the spectrum. (From Haemmerli *et al.*, 1982a. Reprinted by permission of the publisher, The Electrochemical Society, Inc.)

IV. Measuring Systems

A. REFERENCE ELECTRODES

As in any other potentiometric measurement, half of the signal originates from the reference electrode (including the liquid junction). It is, therefore, necessary to pay equal attention to the design factors of the whole measuring system. Of course, it is possible to use a conventional macro reference electrode. However, in that case the advantage of small size of the ISFET is partially lost.

A conventional reference microelectrode sterilizable with ethylene oxide (McBride and Janata, 1978) was designed for *in vivo* studies with ISFETs. A different approach is to build a reference electrode directly on the transistor chip. Comte and Janata (1978) built a small cavity on the transistor chip, which was then filled with a buffer and connected to the solution through a capillary liquid junction (Fig. 28). Because of the small diameter of the capillary junction the buffer solution maintained its composition for several hours. The advantage of this design was the differential mode of measurement, which provided thermal as well as common noise compensation.

There have been two other reports of reference gate ISFETs (Nakajima et al., 1982; Tahara et al., 1982). Both groups used a thin polymer film, either plasma-polymerized polystyrene or ion beam-sputtered Teflon. Both materials show little or no dependence of the drain current on pH. On this basis they were claimed to be suitable sources of reference potential. Unfortunately, the lack of pH dependence does not imply suitability of a device as a reference electrode. In light of the discussion in Section I,B it can be seen that these materials would fall into the category of poor ion-selective membranes or a poor liquid junction, and as such would be subject to nonspecific interferences from small inorganic ions as well as from various surface-active compounds. Although the desire to build a completely solid state reference ISFET is understandable, such a design cannot violate the basic electrochemical rules governing such devices.

There is another possible way to obtain a reference signal. We must realize that for the operation of a CHEMFET we need a signal return, but not necessarily a stable reference potential. Of course, if such a contact changes its potential with changes in solution composition in an unpredictable way, the output of the CHEMFET cannot be uniquely related to the activity of the species of interest. Suppose, however, that the measurement is made transiently, as in *flow injection analysis*. What is then required is a "reference" potential that is stable for the duration of the measurement, which is typically 10–15 s. Such a "reference" CHEMFET can be obtained, for example, by covering an ordinary ISFET with a layer of uncharged gel. The effect of the gel is to slow the ISFET response by

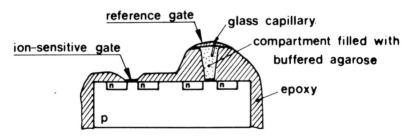

Fig. 28. Reference gate ISFET. (Reprinted with permission from Comte and Janata, 1978.)

increasing the diffusion path. In effect, the gel acts as a liquid junction. If this "slow" ISFET is used in conjunction with an ordinary ISFET, the reference signal is obtained on the basis of the time-resolved response of these two devices.

B. MEASURING CIRCUITS

The CHEMFET can be operated in two modes: constant applied gate voltage and constant drain current. The corresponding circuits are shown

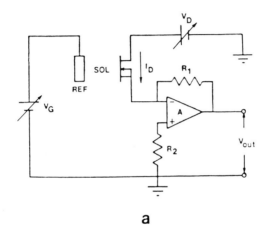

a

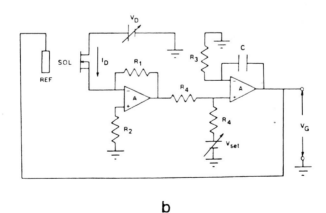

b

Fig. 29. Electrical circuits used in measurement with ISFET at (a) constant applied gate voltage and (b) constant drain current. A, operational amplifier (T1 741C); $R_1 = 1$ kilohms; $R_2 = 470$ ohms; $R_3 = 20$ kilohms; $R_4 = 100$ kilohms; $C = 10$ pF.

in Fig. 29a and b, respectively. The main advantage of operation with a
constant applied reference potential is that several devices in the same
solution and on the same substrate can be run simultaneously. The main
drawback is that the output (drain current) is related to the solution activity of species via ISFET drain current Eq. (19), which lacks the clarity of
the Nernst equation. Also, Eq. (19) is only approximate and cannot be
used to predict the absolute value of the drain current for a given composition of the solution and given applied voltage.

The nonideality of the drain current equation and the implicit nature of
the drain current–activity relationship can be avoided by operating the
device in so-called feedback mode (Fig. 29b). In this mode the drain
current is kept constant by applying a compensating feedback voltage V_{FB}
to the reference electrode. The output of the ISFET measuring circuit is
then read directly in millivolts and is related to the solution activity by the
Nernst equation. The disadvantage of this circuit is that only one device
on a common substrate can be operated in this mode. Another problem is
related to the fact that the response time to an applied gate voltage is
relatively slow ($\sim$ milliseconds for polymeric gate ISFETs), which can
lead to oscillations in the gate feedback circuit. If the interface between
the CHEMFET and the solution is polarized, or if the interface between
the ISFET membrane and the insulator is capacitive, this is the only
correct mode of operation.

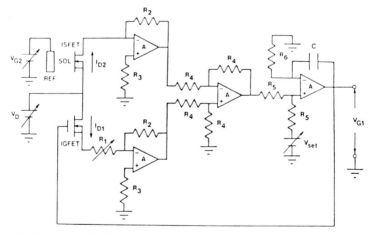

Fig. 30. Differential feedback circuit. A, operational amplifier (T1 741C); R_1 = 0–1
kilohms; R_2 = 1 kilohm ; R_3 = 470 ohms; R_4 = 10 kilohms; R_5 = 100 kilohms; R_6 = 20
kilohms; C = 10 pF.

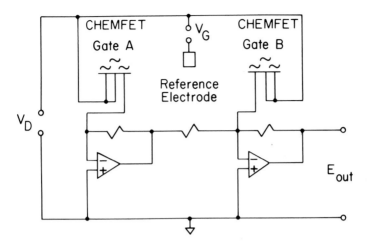

Fig. 31. Differential drain current meter.

Another possibility, although less rigorous than the previous one, combines the best features of the circuits with constant drain current and constant applied gate voltage. This is essentially a differential current follower (Fig. 30), in which the difference between the two drain currents

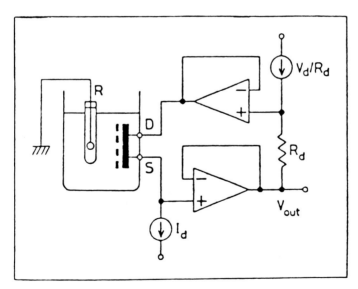

Fig. 32. Measuring circuit for ISFET at constant V_D and I_D. (Reprinted with permission from Nakajima *et al.*, 1980.)

is converted into a voltage, which is then fed into the IGFET. The disadvantage of this circuit is that it requires one IGFET for each CHEMFET. However, this is really no problem, given the large-scale circuit integration that is routinely achieved in modern integrated circuit fabrication. Clearly, this configuration allows multiple CHEMFET operation even with common-substrate devices. The compensating IGFET has a time constant typically about 10^{-7} s, which means that there is no problem

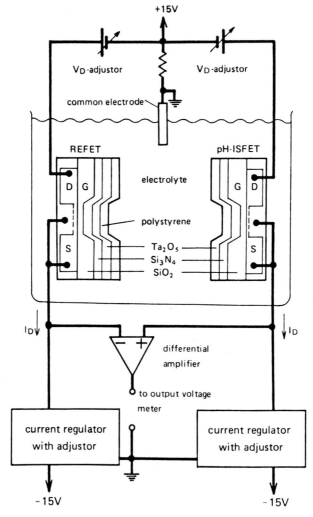

Fig. 33. Current-driven differential circuit. (Reprinted with permission from Tahara *et al.*, 1982.)

with feedback loop oscillations. Another major advantage is that the location of the IGFET on the same chip provides excellent temperature, common noise, and light compensation. Despite the fact that the operating point of the CHEMFET changes with the composition of the solution, this is the circuit of choice for most electroanalytical applications.

Another circuit for differential readout of current between two CHEMFETs is shown in Fig. 31. This circuit is equivalent to the IGFET operated at a constant applied reference electrode voltage (Fig. 29a). The output is current and must be calibrated with respect to the concentration of the measured species. Again, this mode allows multisensor operation because several ISFETs can be measured against a single reference ISFET.

In this survey of measuring circuits we are trying to cover the basic modes of operation. Of course, the actual implementation of these circuits will differ in detail. Matsuo and colleagues (Nakajima *et al.*, 1980) have used a circuit, shown in Fig. 32, in which the common reference electrode is grounded and the CHEMFET is floating. The drain current is maintained constant. This allows simultaneous operation of several devices that are not on a common substrate. Tahara *et al.* (1982) have used the circuit shown in Fig. 33. In this circuit the solution side of the transistor is

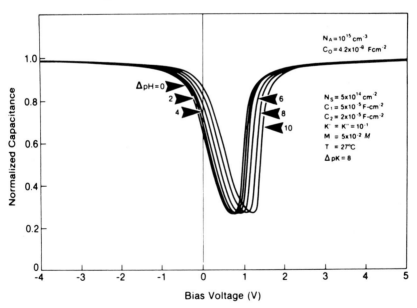

Fig. 34. Experimental capacitance–voltage curves obtained from electrolyte–SiO_2–Si. Temperature, 27°C; oxide capacitance, 4.5×10^{-6} F/cm, measured at 10 Hz. (From Sin and Cobbold, 1979. © 1979 IEEE.)

kept at the potential of the grounded reference electrode while the pair of transistors is floating. The drain current I_D is maintained constant by adjusting the respective drain-to-source voltages.

A considerable amount of work related to the development of ISFETs was done by using an impedance bridge method of capacitance measurement (Buck and Hackleman, 1977; Siu and Cobbold, 1979; Vlasov and Bratov, 1981). There is one major advantage in using capacitance–voltage curves: one does not have to build transistors in order to study some of the basic properties of the gate electrolyte structure. Furthermore, the size of the specimen can be optimized to suit the characteristics of the bridge used for the work. This approach is not, however, suitable for most electroanalytical applications. A set of experimental curves obtained with $Si–SiO_2$–electrolyte is shown in Fig. 34. The shift of the flat-band voltage V_{FB} is due to the change of the interfacial potential and is interpreted in exactly the same way as the change in, for example, feedback voltage of an ordinary ISFET. However, the change of the shape of the $C–V$ curves in Fig. 34 indicates that some much more complex process is taking place in these devices.

V.   Special CHEMFETs

In this section we shall discuss CHEMFETs in which an additional feature has been incorporated into the gate for the purpose of improving the performance or achieving some special function.

A.   Enzymatically Sensitive Field Effect Transistor (ENFET)

The earliest reported device of this type (Danielsson *et al.*, 1979) was a hydrogen-sensitive MOSFET with a palladium metal gate, which is also sensitive to gaseous ammonia. This device was placed in a cuplike compartment in which the gaseous ammonia was generated from the sample solution by an enzymatic reaction (Fig. 35). It is perhaps a question of semantics whether this arrangement should be called a ''palladium MOSFET used for monitoring enzymatically produced ammonia'' or an enzymatic transistor. Nevertheless, the small size and solid state construction of the device clearly show some potential advantages over conventional electrodes, namely baseline stability and speed of response.

An enzymatic transistor sensitive to penicillin has been reported (Caras and Janata, 1980). In this device (Fig. 36) the penicillinase layer was placed directly over the gate of a pH ISFET. The local change of pH

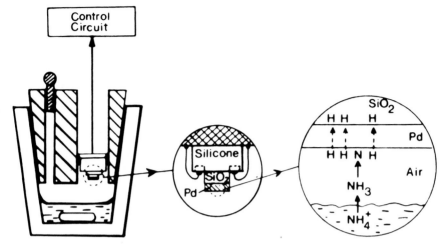

Fig. 35. Schematic representation of the Pd–MOS transistor. The NH_3-sensitive device was mounted in a small glass vessel as shown at the left in the figure. Encircled in the middle is a cross section of the structure. To the right is illustrated the dissociation of NH_3 on the surface of the palladium layer and the diffusion of hydrogen atoms through this layer into the SiO_2. (Reprinted with permission from Danielsson *et al.*, 1979.)

in layer (6), which results from the reaction is monitored by using a

differential current meter (Fig. 31). It is advantageous to use an identical transistor without penicillinase as a reference. This provides all the benefits of differential current measurement, as mentioned earlier, as well as compensation for any pH changes in the ambient solution. A major advantage of the ENFET over an equivalent enzyme electrode is its small size; the amount of enzyme required to make a sensor is extremely small (typically 10^{-4} IU). Furthermore, no retaining membrane is needed to hold the enzyme gel, which results in a shorter response time.

Recently, enzymatically coupled pH ISFETs have been used for measurement of urea and acetylcholine (Miyhara *et al.*, 1983). The same group previously reported a similar device in which trypsin was used to test for α-N-benzoyl-DL-arginine *p*-nitroanilide (BAPNA). In all these devices the active enzyme was immobilized only at the surface of a supporting membrane, which in both cases consisted of a mixture of triacetylcel-

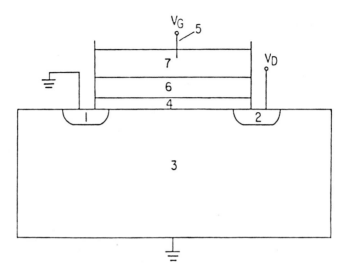

Fig. 36. Schematic diagram of ENFET. (1) drain; (2) source; (3) substrate; (4) insulator; (5) reference electrode; (6) albumin membrane (with or without penicillinase); (7) solution. (Reprinted with permission from Caras and Janata, 1980. Copyright 1980 American Chemical Society.)

lulose and 1,8-diamino-4-aminomethyl octane cross-linked with glutaraldehyde. These devices were used in small volumes of the substrate solution. The surface-immobilized enzyme affected conversion of the substrate in the solution, thus causing a change in the bulk pH, which was detected by the pH ISFETs. The time response of the trypsin device was approximately 6 h, whereas the urea and acetylcholine devices responded in several minutes. It can be speculated that the pH ISFET responded to the bulk changes of pH rather than to pH changes inside the membrane layer (Caras and Janata, 1980), and that similar results would have been obtained by immobilizing the enzyme on any other surface inside the reaction vessel. For this reason, these devices cannot be classified as ENFETs. Hanazato and Shono (1983) reported a true ENFET, which is again based on measurements of local changes of pH due to the enzymatically catalyzed formation of gluconic acid from glucose. They constructed one sensor utilizing purified glucose oxidase in mixture with bovine serum albumin and water-soluble photosensitive polymer. A second type of sensor utilized fermentation microorganisms entrapped in agar gel. A response time of approximately 1 min and a dynamic range of approximately 10–600 mg/l glucose were obtained. In both sensors the active ISFET gate was measured differentially with respect to an identical reference pH ISFET, using a Pt wire as the solution contact. It is noteworthy that this sensor was used under flow-through conditions, which could not be the case for the urea or acetylcholine determinations described above.

B. ELECTROSTATICALLY PROTECTED ISFET

The high input impedance of ISFETs (10^{14} ohms) makes these devices vulnerable to electrostatic damage, particularly during the encapsulation. Static charge can be readily transferred from the operator to the device surface and several hundred volts of static electricity can easily be coupled to the surface, corresponding to a typical field strength in the gate insulator of the order of 10^7 V/cm. These large fields can cause dielectric breakdown and/or shifts in the threshold voltage and subsequent threshold voltage instabilities. Both problems have been observed in ISFETs (Janata and Huber, 1980) and are particularly troublesome in a low-humidity environment. This problem can be avoided to a large extent by increasing the thickness of both silicon dioxide and silicon nitride to 800 Å each. It is also advisable to avoid unnecessary handling of dry devices by an ungrounded operator. While the first solution decreases the sensitivity of the device somewhat, the second one is simply inconvenient. As a

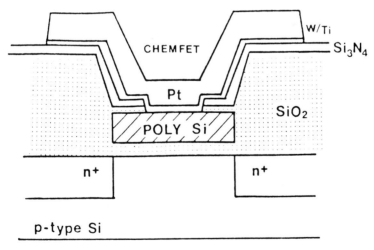

Fig. 37. Cross section of resistively coupled ISFET. The ion-selective membrane (not shown) is applied over the Pt layer. (Reprinted with permission from Smith *et al.*, 1983.)

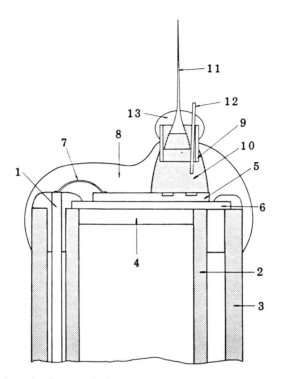

Fig. 38. Schematic diagram of micro ISFET. (1) insulated copper wire; (2) 3-mm-o.d. glass tube; (3) 6-mm-o.d. glass tube; (4) Devcon 5-min epoxy; (5) semiconductor chip; (6) KOVAR substrate; (7) Al–1% Si wire; (8) epoxy (EPON 825 + Jeffamine D230); (9) glass tube; (10) buffered gel; (11) ISM tip; (12) pressure-equalizing capillary; (13) sealing wax. (Reprinted with permission from Haemmerli *et al.*, 1980. Copyright 1980 American Chemical Society.)

solution to this problem the ISFETs were provided with protective circuitry (Smith *et al.*, 1983). Although this circuitry somewhat degrades the input characteristics (specifically, it lowers the input impedance), it also allows the devices to be tested and even electrically calibrated before encapsulation.

The diagram of an electrostatically protected ISFET is shown in Fig. 37. The details of fabrication and its operation are described in Chapter 3. The only membrane with which this type of device was tested was a K^+-selective polymeric membrane. The response of this device to the changes of selectivity of K^+ ion in solution was indistinguishable from that of the conventional K^+ ISFET.

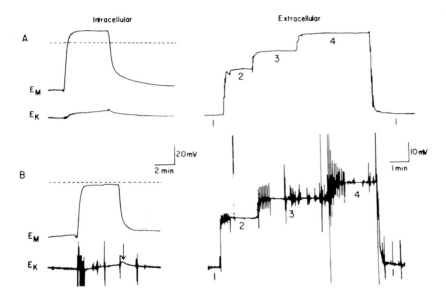

Fig. 39. Physiological application. (A) Micro ISFET recordings; (B) micro ISE recordings. The zero reference for E_M measurements is indicated by the dashed line. For the extracellular recordings, the numbers correspond to the following potassium concentrations in the solution: (1) 8 mM; (2) 100 mM; (3) 239 mM; (4) 470 mM. (Reprinted with permission from Haemmerli *et al.*, 1980. Copyright 1980 American Chemical Society.)

C. MICRO ISFET

Despite their small size, ISFETs are much too big to be used for direct interacellular measurements; there is a size limitation imposed by the necessary encapsulation and by the electrical connections. Yet it is clearly desirable to take advantage of the impedance transformation that these devices offer and to combine them with a micro ISE in which the size of the sensing tip is approximately 1 μm. One arrangement of such a device is shown in Fig. 38. Because of the extremely close coupling of the high-impedance probe tip to the FET preamplifier, the micro ISFET is virtually immune to electromagnetic noise. This feature is illustrated in Fig. 39, in which simple intracellular and extracellular experiments were done with a micro ISFET and a conventional micro ISE. Another improvement gained from this configuration is the shorter response time, which in the case of the K^+ micro ISFET is approximately one-third that of the corresponding micro ISE.

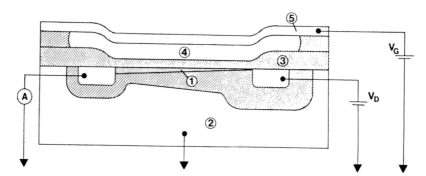

Fig. 40. Schematic diagram of CHEMFET with universally accessible gate. (1) Inversion layer; (2) substrate; (3) insulator; (4) space between insulator and platinum mesh; (5) platinum mesh.

D.   CHEMFET WITH UNIVERSALLY ACCESSIBLE GATE

The development of this device followed from the suspended mesh ISFET (Blackburn and Janata, 1982), which was discussed in Section II,B,2. This CHEMFET is similar in structure to the one with a polyimide gate, except that in this transistor the suspended mesh is made of platinum (Fig. 40) (Blackburn *et al.*, 1983). Although the primary use for this

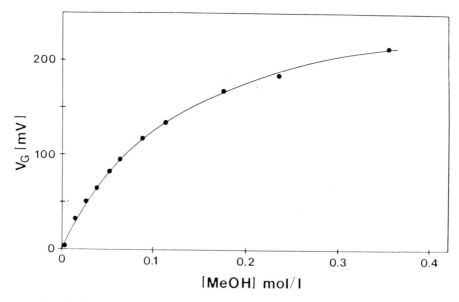

Fig. 41. Response of the suspended metal gate CHEMFET to concentration changes of methanol in toluene.

device is in gas sensing (Chapter 1), it can be used in dielectric liquids as well. The gate (4) can be viewed as an additional "insulator" that is accessible to solutes. The interaction of the solute with the surface of the metal mesh or with the surface of the silicon nitride results in a change of the electric field and hence a change of drain current. An example of such an interaction is shown in Fig. 41, in which the dielectric liquid was toluene and the dipolar solute was methanol. The exploration of this device is still in the early stages. It is, however, obvious that the necessary selectivity can be imparted to this CHEMFET by modification of the interior gate surfaces or by deposition of chemically interactive layers on these surfaces.

VI. Conclusions

Without any doubt the progress made in the development of CHEM-FETs is slow compared to development of the corresponding conventional devices—ISEs. There are several reasons for this: first, there are only a few academic institutions that have the capability of fabricating their own CHEMFET chips. This is in contrast to the ease of fabrication of ion-selective electrodes, which are among the least expensive electroanalytical devices. It is no wonder that a large and growing body of scientific literature exists covering both theoretical and applied aspects of ISEs. Fortunately, much of the knowledge obtained with ISEs is also applicable to CHEMFETs. It is one of the purposes of this review to show that, in principle, CHEMFETs are not different from ISEs and that both operate on the same electrochemical principles. The compatibility of the materials used in construction of CHEMFETs is the major practical problem standing in the way of wide usage of these devices. By deliberate choice both the electrochemical element (the membrane) and the electronic preamplifier (FET chip) are exposed to a very hostile environment, the electrolyte solution. In such a situation, the requirements that are placed on the encapsulation materials are much more stringent than those required for ISEs. Another, related problem pertains to the small size of the chemically sensitive areas, which again presents unique encapsulation problems. It is obvious that truly significant expansion of the development and use of CHEMFETs will not occur until an automatic encapsulation process has been developed and these devices become commercially available at a competitive price.

Inevitably, we need to ask the question: "Is it all worth it?" Theoretically, CHEMFETs can do everything that ISEs can do and more. Over the past 10 years all major theoretical questions about the operation of CHEMFETs have been resolved. It is now clear that the close control over the input capacitance together with the high input impedance of these devices would allow their use with nontraditional membranes. Thus, new types of sensors are likely to be developed. This is particularly significant in view of the fact that the small size of each individual CHEM-FET would enable construction of sensor arrays of acceptable dimensions. Application of these multisensors to chemometrics is expected to bring significant improvement of precision and detection limits of potentiometric sensors.

The preliminary work has also shown that CHEMFETs can be used as new research tools in physiology and medicine, surface adsorption studies, gas detection, and the study of stochastic processes at various interfaces. The very large scale integration silicon technology is much more advanced than the development of new sensors. Once the encapsulation problems are solved the design of sensor packages that will include data acquisition as well as data-processing elements will be feasible.

ACKNOWLEDGMENTS

A considerable amount of the work discussed in this chapter was done at the University of Utah. It is my pleasure to acknowledge the contributions of the following co-workers: G. Blackburn, S. Caras, R. M. Cohen, S. D. Collins, P. A. Comte, A. M. Haemmerli, J. J. Harrow, R. J. Huber, C. C. Johnson, A. Jonkman, M. Levy, P. C. McBride, B. A. McKinley, S. D. Moss, U. Oesch, B. S. Shiramizu, J. Smith, and R. L. Smith.

Financial support throughout the past 10 years was provided by the following institutions: ASULAB SA, Critikon Inc., National Institutes of Health, National Science Foundation, Office of Naval Research, U.S. Air Force, and University of Utah.

REFERENCES

Aizawa, M., Kato, S., and Suzuki, S. (1977). *J. Membr. Sci.* **2**, 125.
Bergveld, P. (1970). *IEEE Trans. Biomed. Eng.* **BME-19**, 70–71.
Bergveld, P. (1972). *IEEE Trans. Biomed. Eng.* **BME-19**, 342–351.
Blackburn, G., and Janata, J. (1982). *J. Electrochem. Soc.* **129**, 2580–2584.
Blackburn, G., and Janata, J. (1984). *Proc. N.Y. Acad. Sci.* **428**, 286–292.
Blackburn, G., Levy, M., and Janata, J. (1983). *Appl. Phys. Lett.* **43**, 700–701.
Buck, R. P. (1978). *In* "Ion-Selective Electrodes in Analytical Chemistry" (H. Freiser, ed.), Vol. I, pp. 1–141. Plenum, New York.
Buck, R. P. (1982). *IEEE Trans. Electron Devices* **ED-29**, 108–115.
Buck, R. P., and Hackleman, D. E. (1977). *Anal. Chem.* **49**, 2315–2321.
Camman, K. (1978). *Anal. Chem.* **50**, 936–940.
Caras, S., and Janata, J. (1980). *Anal. Chem.* **52**, 1935–1937.
Cohen, R. M., and Janata, J. (1983a). *Thin Solid Films* **109**, 329–338.
Cohen, R. M., and Janata, J. (1983b). *J. Electroanal. Chem.* **151**, 33–39.
Cohen, R. M., and Janata, J. (1983c). *J. Electroanal. Chem.* **151**, 41–45.

Collins, S. D., and Janata, J. (1982). *Anal. Chim. Acta* **136**, 93–99.
Comte, P. A., and Janata, J. (1978). *Anal. Chim. Acta* **101**, 247–252.
Covington, A. K., Harbinson, T. R., and Sibbald, A. (1982). *Anal. Lett., Part A* **15**, 1423–1429.
Danielsson, B., Lundström, I., Mosbach, K., and Stilbert, L. (1979). *Anal. Lett., Part B* **12**, 1189–1199.
Esashi, M., and Matsuo, T. (1978). *IEEE Trans. Biomed. Eng.* **BME-25**, 184–192.
Fjeldy, T. A., and Nagy, K. (1980). *J. Electrochem. Soc.* **127**, 1299–1303.
Fujihira, M., Fukui, M., and Osa, T. (1980). *J. Electroanal. Chem.* **106**, 413–418.
Haemmerli, A., Janata, J., and Brown, H. M. (1980). *Anal. Chem.* **52**, 1179–1182.
Haemmerli, A., Janata, J., and Brophy, J. J. (1982a). *J. Electrochem. Soc.* **129**, 2306–2313.
Haemmerli, A., Janata, J., and Brown, H. M. (1982b). *Sens. Actuators* **3**, 149–158.
Haemmerli, A., Janata, J., and Brown, H. M. (1982c). *Anal. Chim. Acta* **144**, 115–121.
Hanazato, Y., and Shono, S. (1983). *Proc. Int. Meet. Chem. Sens., 1983*, pp. 513–518.
Janata, J. (1975). *J. Am. Chem. Soc.* **97**, 2914–2916.
Janata, J., and Huber, R. J. (1980). *In* "Ion-Selective Electrodes in Analytical Chemistry" (H. Freiser, ed.), Vol. II, pp. 107–174. Plenum, New York.
Janata, J., and Moss, S. D. (1976). *Biomed. Eng.* **11**, 241–245.
Joshi, K. M., and Parsons, R. (1961). *Electrochim. Acta* **4**, 129–140.
Koryta, J., Vanysek, P., and Brezina, M. (1977). *J. Electroanal. Chem.* **75**, 211–228.
Lauks, I. (1981). *Sens. Actuators* **1**, 261–288.
McBride, P. T., and Janata, J. (1978). *J. Bioeng.* **2**, 459–462.
McBride, P. T., Janata, J., Comte, P. A., Moss, S. D., and Johnson, C. C. (1978). *Anal. Chim. Acta* **101**, 239–245.
McKinley, B. A., Wong, K. C., Janata, J., Jordan, W. S., and Westenskow, D. R. (1981). *Crit. Care Med.* **9**, 333–339.
Matsuo, T., and Wise, K. D. (1974). *IEEE Trans. Biomed. Eng.* **BME-21**, 485–487.
Matsuo, T., Esashi, M., and Inuma, K. (1971). *Dig. Jt. Meet. Tohoku Sect. IEEEJ.*
Miyhara, Y., Shiokawa, S., Moriizumi, T., Matsuoka, H., Karube, I., and Suzuki, S. (1982). *Proc. Sens. Symp., 2nd, 1982*, pp. 91–95.
Miyhara, Y., Matsu, F., and Moriizumi, T. (1983). *Proc. Int. Meet. Chem. Sens., 1983*, pp. 501–506.
Mohilner, D. M. (1966). *In* "Electroanalytical Chemistry" (A. J. Bard, ed.), Vol. I, p. 241. Dekker, New York.
Moss, S. D., Janata, J., and Johnson, C. C. (1975). *Anal. Chem.* **47**, 2238–2243.
Nakajima, H., Esashi, M., and Matsuo, T. (1980). *Nippon Kagaku Kaishi*, pp. 1499–1508.
Nakajima, H., Esashi, M., and Matsuo, T. (1982). *J. Electrochem. Soc.* **129**, 141–143.
Oesch, U., Caras, S., and Janata, J. (1981). *Anal. Chem.* **53**, 1983–1986.
Sanada, Y., Akiyama, T., Ujihira, Y., and Niki, E. (1982). *Fresenius Z. Anal. Chem.* **312**, 526–529.
Sears, F. W. (1953). "An Introduction to Thermodynamics." Addison-Wesley, Reading, Massachusetts.
Shiramizu, B., Janata, J., and Moss, S. D. (1979). *Anal. Chim. Acta* **108**, 161–167.
Siu, W. M., and Cobbold, R. S. C. (1979). *IEEE Trans. Electron Devices* **ED-26**, 1805–1815.
Smith, R. L., Janata J., and Huber, R. J. (1980). *J. Electrochem. Soc.* **127**, 1599–1603.
Smith, R. L., Huber, R. J., and Janata, J. (1984). *Sens. Actuators* **5**, 127–136.
Solsky, R. L., and Rechnitz, G. A. (1981). *Anal. Chim. Acta* **123**, 135–141.
Sze, S. M. (1969). "Physics of Semiconductor Devices." Wiley, New York.
Tahara, S., Yoshii, M., and Oka, S. (1982). *Chem. Lett.*, pp. 307–310.

Topich, J. A., Fung, C., Wong, A., and Mirtich, M. J. (1978). *Ext. Abstr., Spring Meet.—Electrochem. Soc.,* Abstract 85.

Vadasz, L., and Grove, A. S. (1966). *IEEE Trans. Electron Devices* **ED-13,** 863–966.

Vlasov, Y. G., and Bratov, A. V. (1981). *Sov. Electrochem. (Engl. Transl.)* **17,** 493–497.

Wen, C. C., Lauks, I., and Zemel, J. N. (1980). *Thin Solid Films* **70,** 333–340.

Zemel, J. N. (1975). *Anal. Chem.* **47,** 224A–268A.

3

Fabrication of Solid State Chemical Sensors

ROBERT J. HUBER

DEPARTMENT OF ELECTRICAL ENGINEERING
THE UNIVERSITY OF UTAH
SALT LAKE CITY, UTAH

I. Introduction

A. THE SEMICONDUCTOR FIELD EFFECT

Chemically sensitive field effect transistors (CHEMFETs) are attracting increasing attention from workers interested in their unique analytical possibilities. Of the several theoretically possible types of chemically sensitive semiconductor devices (CSSDs) that have been discussed in the literature (Zemel, 1975), to date only those that utilize the semiconductor "field effect" (Many *et al.*, 1965) have proved to be practical. To be even more specific, the only structure in which the field effect has proved to be

119

useful for building CSSDs is the silicon–silicon dioxide system. It is this system that is used to make the currently successful CHEMFETs. The very extensive technology required to exploit this system has, fortunately, been developed by the electronics industry.

It is sometimes difficult for someone not intimately connected with the fabrication of silicon devices to fully appreciate the depth and breadth of the technology required. In fact, the most practical method of fabrication of such devices makes use of the same facilities required to make large-scale integrated circuits. Anything less will certainly lead to unsatisfactory devices, even for special research work.

In the material that follows enough of this technology is described that a worker can gain a sufficient appreciation of it to at least assess the merits of a particular approach. It is assumed that the reader is familiar with the basic elements of semiconductor band theory. Adequate introductory treatments of the required theory can be found in standard texts (Grove, 1967; Muller and Kamins, 1977). A summary of this theory is given in Chapter 1 by Lundström and Svensson.

The d.c. field effect refers to the steady-state changes in surface conductance induced by electrostatic fields (Many *et al.*, 1965). In simple physical terms, it is due to charge induced on or near the semiconductor surface by an external electric field normal to the surface. Changes in this normal field result in changes in the surface charge density. If this charge (or a fraction of it) is mobile, then induced changes in the normal electric field result in measurable changes in the electrical conductivity of the surface layer. Conversely, by measuring changes in surface conductivity one can deduce changes in the normal electric field. Then, if the details of the structure are known—that is, the capacitance per unit area of the insulator layer on the surface—changes in surface conductivity can be used as a measure of changes in the voltage across the surface capacitance. The insulated gate field effect transistor (IGFET) depends on changes in the surface conductivity induced by changes in an externally applied voltage. The CHEMFET depends on changes in the surface conductance resulting from changes in an electrochemically induced voltage.

For either the IGFET or the CHEMFET to function properly, at least part of the surface charge induced by the normal electric field must be mobile, that is, free to move under the influence of an electric field parallel to the surface. Practically speaking, this places severe restrictions on the materials used to construct the device. In general, whenever a crystal lattice is terminated (i.e., the surface) there will be a large number of allowed electron energy levels within the forbidden band at the surface. Electrons that occupy these levels are often not mobile, but they do serve to terminate the normal electric field. Changes in the applied normal electric field then result in changes in the surface charge density, but not necessarily changes in surface conductivity. Usable field effect transistors can be built only with a material in which these surface states do not easily exchange charge with their surroundings. At present, there is only one really practical system: the silicon–silicon dioxide system. If a properly cleaned silicon surface is oxidized at high temperature in an atmosphere of oxygen, water vapor, or a mixture of the two, the expected surface states are effectively removed from the system. The details of the physics of this system are not fully understood at present, but the practical effect is that, with present technology, no other semiconductor material but silicon covered with thermally grown silicon dioxide can be used to build useful IGFET devices. There is substantial continuing research to develop a deposited insulator coating on the surface of a nonsilicon semiconductor, such as gallium arsenide, which would make a good IGFET, but so far with notable lack of success.

Another practical constraint on the fabrication of these devices is the required cleanliness. Many chemical contaminants induce charges in the surface of the silicon in the IGFET. For example, sodium often acts as a positive charge when it is incorporated in silicon dioxide. It is also mobile enough to migrate through the thin layers found in these devices in short periods of time (minutes). Since the density of the charge layers induced in the surface of an IGFET is of the order of 1×10^{11} cm^{-2}, sodium contamination of this order of magnitude in the oxide can make an IGFET useless. The result is the requirement for very elaborate contamination control methods during fabrication.

The CHEMFET is an insulated gate field effect transistor with the normal conductive gate electrode replaced by some structure that generates a gate voltage by electrochemical means. In principle, any structure that translates a chemical concentration into a voltage can be used to make a CHEMFET. In some cases, the interaction between the gate insulator itself and a solution results in a CHEMFET. A notable example is the pH-sensitive device. A CHEMFET with a layer of silicon nitride (Matsuo and Wise, 1974; McBride *et al.*, 1978) gives a very good response to changes in pH even with no additional gate structure. Other ions in solution may require a more complex gate structure.

The fabrication of CHEMFETs should not be undertaken lightly. The basic concepts of the device have been recognized for some time, but the successful implementation is relatively recent (Janata and Moss, 1976). Production of devices with predictable characteristics depends on the full range of technology that has been developed by the electronics industry for the fabrication of large-scale integrated (LSI) circuits. In fact, CHEMFET fabrication can be considered just a variation of near-standard microelectronics processing, requiring electronic-grade materials (of which more will be said later in this chapter), high-resolution photolithography equipment, and conditions of cleanliness equal to those of LSI facilities. The resolution of the photolithography employed in the fabrication of single CHEMFETs need not be as fine as that for very large scale circuits (presently about 2 μm), but there are emerging techniques that will couple chemically sensitive devices and considerable amounts of signal processing on the same silicon chip. Full exploitation of this technology will remove the distinction between the requirements for CHEMFETs and more conventional integrated circuits. Proper fabrication facilities require large amounts of expensive equipment; a common mistake is underestimation of the amount of capital equipment required.

B. THE FIELD EFFECT TRANSISTOR

The physical principles on which the insulated gate field effect transistor operate have been adequately described elsewhere (Muller and Kamins, 1977; Sze, 1981). The usual result of the analysis is an equation or set of equations that relate certain currents in the device to the voltages applied to the external connections and to the physical structure. For typical devices Eqs. (1) and (2) give the drain current in terms of the applied voltages and device structure.

$$I_D = \mu_n C_{ox} \frac{W}{L} \left[(V_G - V_T) V_D - \frac{V_D^2}{2} \right] \tag{1}$$

$$I_{DSAT} = \mu_n C_{ox} \frac{W}{L} \left(\frac{V_G - V_T}{2} \right)^2 \tag{2}$$

where μ_n is the electron mobility in the surface channel (this is normally about one-third to one-half of the corresponding bulk mobility), C_{ox} is the capacitance per unit area of the gate insulator, W is the width of the channel, L is the length of the channel (source–drain spacing), and V_G, V_T, and V_D are the gate, threshold, and drain voltages.

Equation (1) describes the device in the unsaturated region of operation. In this region, all voltages are such that the surface of the silicon is under conditions of strong inversion everywhere in the channel (Grove, 1967). The channel is the thin surface layer of the silicon between the source and drain that forms the conducting path for the current. Equation (2) describes the saturation region of operation. In this region, the portion of the channel adjacent to the drain is not strongly inverted. Since this condition changes the assumptions used in the derivation of the equations, the functional form of the equations changes. It should be pointed out that Eqs. (1) and (2) are the result of some substantial simplifying assumptions. As such, they are only "qualitatively" correct. They predict the observed shape of the current versus voltage response of the devices. A more accurate description, Eqs. (3) and (4), of the IGFET is obtained from a more realistic (and more complex) analysis (Muller and Kamins, 1977).

$$I_D = \mu_n \frac{W}{L} \left\{ C_{ox} \left(V_G - V_{FB} - 2|\phi_p| - \frac{1}{2} V_D - \frac{1}{2} V_S \right) (V_D - V_S) \right.$$

$$\left. - \frac{2}{3} \sqrt{2\varepsilon_s q N_a} \left[(2|\phi_p| + V_D - V_B)^{3/2} - (2|\phi_p| + V_S - V_B)^{3/2} \right] \right\} \quad (3)$$

$$I_{DSAT} = I_D(V_{DSAT})$$

$$V_{DSAT} = V_G - V_{FB} - 2|\phi_p|$$

$$- \frac{\varepsilon_s q N_a}{C_{ox}^2} \left[\sqrt{1 + \frac{2C_{ox}^2}{\varepsilon_s q N_a} (V_G - V_{FB} - V_B)} - 1 \right] \quad (4)$$

where V_{FB} is the flat-band voltage, ϕ_p is the position (in volts) of the Fermi level relative to the intrinsic Fermi level in the substrate silicon, N_a is the acceptor impurity concentration in the substrate (assumed to be constant with position), V_B is the reverse bias voltage between the source and substrate, V_{DSAT} is the saturation drain voltage, q is the electronic charge, and ε_s is the permittivity of silicon.

Both of the above sets of equations are "one-dimensional" in that they assume that the drain current flows only at the surface of the silicon crystal. For most purposes, they give satisfactory results. However, the electronics industry is now using such small dimensions that even Eqs. (3) and (4) are not sufficient, and it is more appropriate to use two-dimensional numerical solutions to the partial differential equations that describe current flow through the material (Mock, 1973).

C. THE BASIC CHEMFET

The chemically sensitive field effect transistor, CHEMFET, in its most common form, is an insulated gate field effect transistor, IGFET, with the usual gate metal replaced by a more complex chemically sensitive structure (Janata and Huber, 1980). The chemically sensitive components and the transistors are built as part of the same monolithic integrated circuit. Conceptually, one might think of the CHEMFET as a conventional high-input-impedance amplifier attached to an ion-selective electrode measuring device by a "very short" piece of wire. Commonly, this wire has "zero length," but in some of the devices to be described it has a length of a few micrometers.

The dimensions of the CHEMFET are generally large enough that the simple one-dimensional model [Eqs. (3) and (4) above] is adequate. Furthermore, if the CHEMFET is operated in the constant-current mode (Janata and Huber, 1980) there is no need for the more accurate model and Eqs. (1) and (2) can be used. The chemically sensitive structure is placed in series with the gate in such a way that chemically generated potentials are applied to the gate (Janata and Huber, 1980). Equations (1) and (2) can be modified to explicitly display the dependence of drain current on chemical activity, as shown in Eqs. (5) and (6). Equation (7) is the usual equation for the threshold voltage of an IGFET modified for the case of the CHEMFET.

$$I_D = \mu_n C_{ox} \frac{W}{L} V_D \left(V_G - V_T^* \pm \frac{RT}{Z^i F} \ln a_2^i - E_{ref} - \frac{V_D}{2} \right) \tag{5}$$

$$I_{DSAT} = \mu_n C_{ox} \frac{W}{2L} \left(V_G - V_T^* \pm \frac{RT}{Z^i F} \ln a_2^i - E_{ref} \right)^2 \tag{6}$$

$$V_T^* = -\Delta\phi_{cont} - E_0^i - \frac{Q_{ss}}{C_{ox}} + 2\phi_F - \frac{Q_B}{C_{ox}} \tag{7}$$

where a is the activity of the ion, E_{ref} is the reference electrode potential, z^i is the number of elementary charges, F is the Faraday constant, $\Delta\phi_{cont}$ is the contact potential between the semiconductor and metal, Q_{ss} is the fixed charge in the gate oxide, and Q_B is the charge in the silicon surface space charge region.

D. BIPOLAR TRANSISTORS VERSUS IGFETs AS CHEMFETs

The bipolar transistor (Grove, 1967) has been used in semiconductor electronics longer than has the IGFET. It might be asked why monolithic integrated circuits combining bipolar transistors and chemically sensitive structures have not been built. If this were possible, several of the problems with the current CHEMFET devices would be avoided. In particular, the serious sensitivity to static electric charge would not be present. The gate of the IGFET is one plate of a capacitor with a dielectric thickness of 1000 Å or less. The dielectric strength of silicon dioxide is about 10^7 V/cm. A gate voltage of about 100 V, which can easily be obtained from static charge during handling, is sufficient to destroy the devices. The bipolar transistor, however, is a current-actuated device, whereas the IGFET is a voltage-actuated device. If used in a chemically sensitive integrated circuit based on bipolar transistors, the chemically sensitive structures might be required to supply more current than would generally be possible. In other words, the input impedance of the bipolar transistor is generally too low to be driven by many of the chemical systems of interest.

E. JUNCTION FIELD EFFECT TRANSISTOR VERSUS IGFETs AS CHEMFETs

The junction field effect transistor (JFET) is in many important ways similar to an insulated gate FET. A discussion of the operating principles and the equations describing the current–voltage characteristics of this device may be found in many standard texts (Grove, 1967; Muller and Kamins, 1977). This theory was first developed by Shockley (1952). Like the IGFET, the JFET is a voltage-actuated device. In this device, the conductivity of a channel is modulated by a reverse bias applied to the *pn* junctions that define the channel. The d.c. input impedance is that of the reverse-biased junction. This can be very high, of the order of (10^{12}) ohms for a well-prepared junction with a small area. There are many chemical systems of interest for which this would be an acceptable impedance to drive, and it may be practical to build monolithic chemical sensing systems based on the JFET. In such a device, an ohmic contact to the gate junction could be made with any of several materials such as aluminum or platinum silicide (Poate and Tisone, 1974). The chemically sensitive part of the device would be in electrical contact with the metal. By properly adjusting the voltage on a reference electrode so that the voltage applied to the gate *pn* junction is kept small (generally near zero or reverse-biased), it is, in principle, possible to make an integrated chemical sensor with a transconductance of the same magnitude as that of the CHEMFET.

The later developments in CHEMFETs described in this chapter include an electrostatically protected CHEMFET. In this device, a *pn* junction is intentionally placed in parallel with the insulated gate. This, of course, degrades the normally very high input impedance of the IGFET, but it has been found experimentally to be acceptable in many cases. As the technology of CHEMFET fabrication develops, it is possible that use of a JFET as the input semiconductor device on a monolithic chemically sensitive integrated circuit will become commonplace.

II. General Fabrication Procedure

A. THE SILICON PLANAR PROCESS

By far the most powerful procedure for the fabrication of semiconductor devices is the general collection of steps known as the planar process.

It is based on the use of very high resolution photolithography to define selected areas of silicon for the addition of donor and acceptor impurity atoms as well as the other necessary structures such as metal interconnections. Patterns to be built into silicon are first made as high-resolution, high-contrast photoplates. These plates will be the same size as the finished device if they are to be transferred to the silicon by contact printing, or several times larger if proper optical projection equipment is available.

Donor and acceptor impurities are added to the silicon in either of two ways. The oldest and most common method is thermal diffusion at an elevated temperature, making use of the fact that thin layers of SiO_2 will adequately "mask" the diffusion of appropriately selected impurities. The commonly used donor and acceptor atoms generally have much lower diffusivity in silicon dioxide than in silicon. The other method, which is rapidly becoming the standard method in many facilities, is ion implantation. This method uses a small positive-ion accelerator to physically implant the impurity atoms into the silicon. Both of these methods are used in the process described in this chapter.

Many other materials used in the planar process are deposited on the surface of the devices as continuous layers by chemical or physical vapor deposition processes, then patterned by additional photolithographic steps. Following is a short description of each of the basic steps used in our procedures. The actual fabrication sequence for an n-channel metal gate process is then discussed.

1. Silicon Wafers: Manufacture, Characterization, Sources, etc.

The silicon is used in the form of circular wafers cut from high-purity, mechanically near-perfect "boules" usually grown by the Czochralski (Ravi, 1981) process. They are dislocation-free and their impurity content is tightly controlled. The crystallographic orientation of the surface of the wafer is important. A short review of the essential aspects of silicon crystal growth and wafer preparation has been given by Pearce (1983). The process described here uses p-type wafers having a $\langle 100 \rangle$ surface and a resistivity of 5 to 10 ohm-cm. Such material is readily available commercially.

2. Oxidation

Many times during fabrication, a thin layer of SiO_2 is grown on the wafer by what is known as thermal oxidation. This is done by placing the wafers in a high-temperature resistively heated furnace, as shown in Fig. 1, in an atmosphere of either pure oxygen or a water vapor and oxygen mixture. The kinetics of this process are well known (Deal and Grove,

Fig. 1. Semiconductor diffusion furnace suitable for laboratory use. Three separate temperature controllers control the temperature in the center and end zones of the furnace. ("Mini-Brute" photograph courtesy of Termco Products Corporation, Orange, California.)

1965), and layers of closely controlled thickness are readily obtainable. The most critical part of the oxidation procedure is the purity of the environment. The furnaces are lined with high-purity fused silica tubes and all reagents used are of carefully controlled purity. All carriers used in these high-temperature environments are either fused silica of semiconductor quality or high-purity polycrystalline silicon. Trace amounts of heavy metals are particularly troublesome as they result in very effective recombination centers in the silicon. Some oxidation methods call for the addition of chlorine to the furnace atmosphere to control contaminants (Rohatgi et al., 1979). Chlorine is commonly added either as anhydrous HCl or as some other vapor such as trichloroethylene added directly to the gas stream. The oxide layer grown under these conditions is amorphous. Any inadvertent crystallization of this oxide layer renders it useless as a diffusion mask. Acceptable procedures for growing a predetermined thickness of oxide are well established in the electronics industry (Katz, 1983).

3. Diffusion

The necessary amounts of dopants that form the source and drain regions of the field effect transistors are added by solid state diffusion into regions defined by photolithography. Usually, openings are etched through the layer of SiO_2 that covers the surface and the dopants are diffused or implanted through these openings. The dopants are added to the substrate either by chemical means or by ion implantation in a step generally known as predeposition. The chemical processes, which require less costly equipment, have been used much longer than has ion implantation. The chemical processes provide much less precision in controlling the amount of dopant added. Because of this, ion implant is rapidly becoming the preferred method. However, for the process described here, the source and drain diffusions involved chemical predeposition of phosphorus. Following the predeposition step, the dopants are driven into the crystal by solid state diffusion. An introductory treatment of the details of this process is given by Grove (1967). In the process used here to make CHEMFETs, the junction depth of the diffused regions is about 1.5 μm.

4. Image-Forming Processes

Common to all planar processes are the photo steps that transfer the patterns which define the device being built onto the silicon wafer. A lengthy review of the current state of this technology has been given by Elliott (1982). The technology for this process is developing very rapidly as device dimensions are being reduced. Direct contact printing of the mask onto the wafer is the oldest and easiest to implement, but it has several disadvantages. Images may be projected onto the wafer with a 1:1 magnification ratio, they may be projected at a reduced size onto the wafer, or they may be "written" directly onto a sensitized wafer with a scanned electron beam. Some investigators have even used soft X-rays to project the images onto the wafer.

Because of the substantially lower cost of the equipment required, contact printing is the method most often employed if the devices or circuits are not very large and if the minimum dimensions are not less than about 5 μm. Figure 2 shows a standard mask aligner used for aligning a mask to the patterns already present on the wafer and making the contact exposure. The most serious problem with contact printing is the mechanical damage to the mask caused by the actual contact. Defects introduced into a mask in this way will be reproduced in all succeeding wafers that use the same mask, resulting in lower yields of fully functional devices. Since masks are relatively costly, the decision to use or not to use contact printing methods is largely an economic one. Because most CHEMFETs

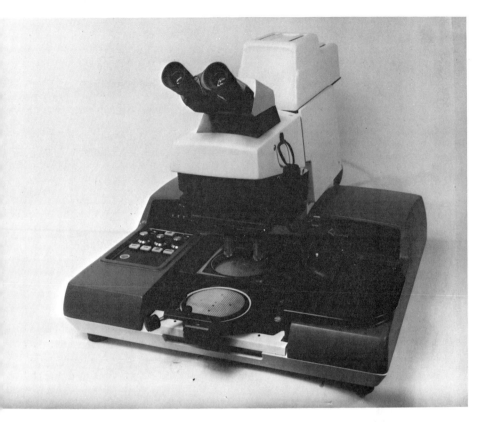

Fig. 2. Mask alignment and exposure system. Such a system is equipped with precise mechanical positioning and a high-resolution split-field microscope. (Photograph courtesy of Quintel Corporation, San Jose, California.)

are small and have rather large dimensions, 5 μm or greater, contact printing is the most frequently used method. The cost of projection equipment could be justified only if a particular CHEMFET design was in mass production or if a very large device was necessary. In this context, a large device may require a silicon area of 4.0×10^7 μm^2. The next several paragraphs outline the basic sequence of steps in the contact printing process; the necessary modifications for use with a projection process can easily be made.

Wafers are cleaned both mechanically and chemically to remove all traces of contamination. Mechanical cleaning is done by a specially designed apparatus with brushes of inert material, or by high-pressure jets of deionized water. Mechanical cleaning is not used if the level of particulate contamination is low. However, the wafers are always chemically cleaned before any high-temperature operation. Various cleaning solutions are used. Two of the most popular are a mixture of concentrated sulfuric acid and hydrogen peroxide, and a mixture of ammonium hydroxide and hydrogen peroxide followed immediately by a mixture of hydrochloric acid and hydrogen peroxide (Kern and Puotinen, 1970). After exposure to these cleaning solutions, as is true for most other steps involving wet chemistry, the wafers are always given a lengthy rinse in deionized water. The commonly used laboratory cleaning agents containing chromic acid should never be used because of the contamination of the silicon by trace amounts of chromium. The wafers are then dried in a dust-free environment.

Photoresist is usually applied to an oxide surface as soon as possible after the wafer is removed from the oxidizing furnace. Delays that entail exposing it to a humid laboratory atmosphere or process steps that involve immersing the wafers in water result in the oxide surface being hydrated. The layer of water molecules on the surface reduces the photoresist adhesion and allows the subsequent etching solution to penetrate between the photoresist and the oxide, causing serious loss of resolution.

A chemical adhesion promoter is frequently applied to the wafer just prior to application of the photoresist. The most common one is hexamethyldisilazane (HMDS).* Trichlorophenylsilane or other related compounds can also be used. Without an adhesion promoter, the adhesion of the photoresist to a silicon dioxide surface is often insufficient to prevent loss of resolution during subsequent etching steps. If the oxide surface has not absorbed water molecules, adhesion is adequate, but the detrimental layer of water molecules is readily provided by water vapor in the atmosphere unless the relative humidity is very low. Attempts to avoid loss of photoresist adherence by maintaining a very dry laboratory have been only moderately successful.

Photoresist is applied to the wafers by spin coating, producing layers of very uniform thickness. In this process, the wafer is placed on a spindle and held in place by vacuum, then flooded with photoresist and spun about its axis at high speed, usually 2000–5000 revolutions per minute. It is imperative that the coating be done in a dust-free place and that the coated wafers be kept free of any particulate contamination until processed. The usual particle of dust is comparable in size to the patterns that will be printed onto the photoresist. After spin coating, the wafers are baked at about 90°C to eliminate remaining solvent. Care must be taken to avoid temperatures high enough to degrade optical sensitivity.

---

* The use of HMDS as an adhesion promoter in microelectronics fabrication is described by Collins and Deverse (1970).

Exposure of the photoresist in the contact process is done through a high-contrast photomask. Both positive and negative resists are commonly used. Positive resist becomes soluble where exposed and negative resist becomes insoluble where exposed in the appropriate developer. Obviously, the mask requirements are different in each case. When using contact mask exposure methods, it is often best to use a mask–photoresist combination in which most of the mask is transparent (known as a "clear field" mask). This makes it much easier to see the patterns previously etched into the silicon during the alignment. Exposure times are usually adjusted to be a few seconds. Before the actual exposure is made, the masks must be accurately aligned with existing patterns on the wafer; if one is making the initial exposure, the masks should be aligned along a preferred crystallographic direction of the silicon. The individual devices are often cut from the silicon wafer by a scribe and fracture method. This process works best for certain crystallographic orientations.

Tolerances for alignment of the mask to previous patterns vary depending on the dimensions of the device, but must be a small fraction of the minimum feature size of the device. These dimensions are continually being reduced in large-scale integrated circuit designs. At present, mass production processes use alignment tolerances of 0.2 μm across a 100-mm-diameter wafer. The designs for CHEMFETs at the present stage of development need not be that precise, but alignment tolerances near 1 μm may be needed for multisensor designs or for designs that provide some signal-processing circuitry on the same chip. Even these larger tolerances require the use of specialized mask alignment and exposure equipment. In a silicon device fabrication laboratory, the mask alignment equipment may represent one of the largest capital investments.

After exposure, the latent image is developed by either spraying the wafer with developer or immersing it in a container of developer. The actual method depends on the type of photoresist used; manufacturers generally will provide the necessary information. During development, it is necessary to completely remove the resist in the image-forming areas. In most chemical etching steps, even a slight scum of unremoved resist will inhibit the etching process enough to ruin even the largest geometric features. Therefore, after development the wafers must be rinsed thoroughly in a suitable solvent, such as deionized water. Many silicon-processing laboratories find it necessary to include a plasma "descum" operation, exposing the wafers to an oxygen plasma for a short period of time. This step removes a small amount of photoresist everywhere, but the photoresist in the unaffected areas is thick enough that its masking acceptability is not affected.

After development of the image, the wafers are usually baked again at about 120°C to enhance photoresist adherence. Particularly when using wet etches, there is a tendency for the resist to lift at the edges, resulting in loss of pattern resolution. When using plasma etching or reactive ion etching, the need for this postbake is not as great.

5. Etching

There are two general types of etching processes. The simplest uses conventional wet chemistry; the other makes use of low-pressure plasmas and is known as "plasma etching" (Chapman, 1980). Wet etching can readily be done with simple laboratory apparatus, while plasma etching requires complex and expensive equipment. Plasma etching does have several advantages for some particular combinations of materials used in the fabrication of CHEMFETs and is the preferred method, if available, in etching a layer of dense silicon nitride. It is possible to etch silicon nitride with hot (180°C) phosphoric acid (Van Gelder and Hauser, 1967), in which case the masking is done by a layer of silicon dioxide, which in turn has been patterned with a photoresist mask and room-temperature buffered hydrofluoric acid. The same layer of silicon nitride can easily be etched in a CF_4 plasma, using only photoresist as a mask.

In the CHEMFET fabrication described later in this chapter, etching of layers of SiO_2, polycrystalline silicon, silicon nitride, and aluminum is required. All of these can be done with wet chemistry. Figure 3 shows a standard "wet station" equipped with a laminar-flow clean air hood. Laminar-flow hoods are routinely used to reduce particulate contamination (Kilpatrick, 1984). The SiO_2 layers are etched with buffered hydrofluoric acid, but the etch rate of concentrated HF if too high to give reproducible results with the thin films (a few tens to a few hundred nanometers thick) employed in CHEMFETs, so dilute solutions of HF in water or HF solutions buffered with NH_4F are used. Etch rates are generally kept to 100 nm/min or less. Neither of these solutions etches silicon, so the oxide etching process is self-terminating with regard to depth. However, careful control of the etching time is still necessary to prevent the etch from undercutting the edge of the photoresist pattern. Polycrystalline silicon used in the electrostatically protected CHEMFET is patterned by etching in a solution of acetic acid, nitric acid, and hydrofluoric acid. The general procedures are similar to those described for oxide etching. Wet etching of silicon nitride requires more elaborate methods. Silicon nitride, as used by the microelectronics industry, does not have a unique composition. There appears to be a wide range of compositions and properties, depending on the details of the deposition process used to form it. Of most interest here are its chemical properties. Silicon nitride

Fig. 3. Standard wet station with laminar air flow. Stations like this with exhaust are required for etching processes carried out under clean conditions. (Photograph courtesy of Integrated Air Systems, Inc., Valencia, California.)

deposited at high temperatures is very resistant to attack by HF, while nitride deposited at lower temperatures, or nitride deposited by a plasma-aided process, can be etched in buffered HF. Wet etching of the high-temperature nitride can be done with 180°C phosphoric acid (Van Gelder and Hauser, 1967). Immediately following completion of the etch, the wafers must be rinsed thoroughly in deionized water.

6. Water Purity Requirements

Microelectronic fabrication makes use of large quantities of very pure water (Taubenest and Ubersax, 1980). The most common method of purifying water for such use is by deionization. The purity of water used in microelectronic device fabrication is generally monitored by measuring the electrical resistivity, which is normally kept above 14 megohm-cm. The details of a particular installation depend on the chemical composition of the water available at the laboratory site. A typical installation will employ initial filtration to remove particulates, followed by treatment to remove organic contaminants. The water is then given a first-stage purification by reverse osmosis. After this step, the resistivity of the water is of the order of 10^5 ohm-cm. It is further purified by passing through mixed-bed ion exchange columns and then distributed to the laboratory by a continuous-flow system made of carefully selected materials, PVC being the most common. Care must be taken to avoid the build-up of colonies of microorganisms in the system. Most fabrication facilities require the rinse water at the point of use to have a room-temperature resistivity above 10 megohm-cm. A typical system will maintain water with resistivities greater than 14 megohm-cm at room temperature. The result of using water of significantly poorer quality is unstable electrical parameters of the semiconductor devices, caused by waterborne contaminants.

7. Ion Implantation

Ion implantation (Seidel, 1983) is used routinely in device fabrication to deposit the dopant atoms in very precise quantities on or near the surface of the semiconductor crystal. Some advanced processes use it as a predeposition source for a subsequent diffusion step. The alternative and older chemical methods for predeposition are often satisfactory, but in the fabrication of insulated gate field effect devices ion implantation has become an indispensable tool for the control of threshold voltage. The threshold voltage is the gate-to-substrate voltage required for the formation of a conducting channel from the source to the drain. The nature of the silicon–silicon dioxide interface often causes the surface of a lightly doped p-type wafer (the substrate for forming n-channel devices) to invert to n-type. Unless this effect is eliminated, the inactive portion of an n-channel chip will have unwanted current paths and the actual transistors will be turned on even with no applied gate voltage. These problems usually make the device useless. Formation of this detrimental n-type layer can be prevented by ion implantation. Acceptor atoms, usually boron, are implanted into the surface of the silicon to an area density of a few times 10^{12} cm^{-2}. Ion implantation is also used to set the threshold voltage to the desired value.

The ion implanter (Fig. 4) is a small electrostatic positive-ion accelerator, usually with a maximum accelerating voltage of 200 kV. The accelerating voltage is easily adjustable. Beam currents of machines now in use range from microamperes to milliamperes of the species being implanted. The machines are fitted with beam-scanning equipment (either electrical or mechanical), and the ion beam is scanned over the entire wafer, taking great care to achieve uniformity. In standard fabrication methods, no attempt is made to scan only selected areas of the wafer, although this is an active area of research.

At the voltages used, the implantation range of the ions in the semiconductor materials is submicrometer, making it easy to mask portions of the silicon to prevent the ions from reaching it. The ion implantation mask is usually photoresist, but other layers, such as silicon dioxide or metal films, are used. The density of the implanted ions is measured by integration of the beam current. The implanter is designed so that the wafer being implanted is held in an electrically isolated Faraday cup. The current delivered to the wafer by the ion beam is returned to ground through a current integrator so that the charge accumulated by the current integrator is a direct measure of the total ion dose received by the wafer. This method allows close control of the number of dopant atoms in selected areas of the wafer. It is easy to accurately deposit dopant atoms in the silicon in the range of 10^{10} atoms/cm^2, which is several orders of magnitude smaller than the density that can be achieved reproducibly by more conventional chemical means. Following ion implantation, the wafers must be given a high-temperature treatment to anneal the crystalline damage caused by the implantation and to allow the dopants to move onto electrically active sites in the crystal.

8. High-Vacuum Deposition Processes

Ohmic electrical contacts to the silicon must be made, as well as interconnections for the on-chip circuitry. Aluminum is the most widely used metal on silicon devices. It is easy to deposit, adheres well to silicon dioxide, and makes good ohmic contacts. Aluminum can be deposited in

Fig. 4. Ion implanter. (Photograph courtesy of Varian Associates, Inc., Palo Alto, California.)

vacuum by several methods, the most common being filament evaporation, electron beam evaporation, and sputtering. Filament evaporation, in which an aluminum charge is evaporated from an electrically heated tungsten filament, has not proved to be very successful in the fabrication of insulated gate field effect transistors because of contamination of the surfaces of the silicon wafer. Electron beam evaporation is much more satisfactory. In this method, a crucible of aluminum is heated by an intense, high-voltage beam of electrons. The beam is focused on the center of the charge. This greatly reduces contamination because only the electron beam touches the evaporation source. The main problem with this technique is the difficulty in depositing alloys. Use of an aluminum alloy containing 1–2% silicon is very common. Other alloying elements such as copper are also used. Aluminum sputtering (Chapman, 1980) is very widely used, especially in mass production and where an alloy is needed.

In any of the vacuum deposition methods for aluminum, great care is needed to avoid detrimental amounts of water vapor in the residual gas in the vacuum system during evaporation. Aluminum is so active chemically that any residual water vapor results in Al_2O_3 in the metal. Evaporations should be performed at pressures of 10^{-6} torr or less. When sputtering aluminum or its alloys, the vacuum chamber must be evacuated to this pressure before admitting the inert gas, usually argon, used for the plasma. A partial pressure of water vapor of 1×10^{-5} (Hoffman, 1978) results in so much Al_2O_3 in the metal that both its electrical conductivity and its ability to be patterned by high-resolution photolithography are degraded.

9. Chemical Vapor Deposition

Chemical vapor deposition is used at several steps in the process to deposit layers of silicon dioxide, silicon nitride, and polycrystalline silicon (Adams, 1983). These layers are deposited by the gas-phase reactions of several gases. Silicon dioxide is deposited by reacting silane (SiH_4) and oxygen or, occasionally, CO. It is used whenever it is not practical to grow a layer of SiO_2 by oxidation of Si—for example, when the layer is so thick that it would require an excessive amount of time, when the growth of SiO_2 would consume an unacceptable amount of silicon from the wafer, or when a layer of SiO_2 is needed on top of a layer that is not silicon. Silicon nitride is deposited by reacting either silane or silicon tetrachloride and ammonia at high temperatures (900–1000°C). It can also be deposited at low temperatures (400–500°C), using a nitrogen plasma. In all cases, however, trace amounts of oxygen must be avoided. Polycrystalline silicon is most easily deposited with the required uniformity by the thermal decomposition of silane. The most sucessful method is the thermal decomposition of pure silane near 625°C at low pressure (200 millitorr). Because silane is pyrophoric, well-planned safety precautions are necessary when working with it. Chemical vapor deposition is a very specialized field and the reader should consult the literature for details of particular processes.

B.   DESIGN AND LAYOUT CONSIDERATIONS OF CHEMFETs

This section is a discussion of the dual-gate CHEMFET that the University of Utah research group has successfully used during the past few years. The chip contains two identical CHEMFET gates and two identical "normal" metal gate MOSFETs. The two metal gate devices serve to monitor the fabrication process. The design is executed in n-channel metal gate technology.

The choice of n-channel technology was made to take advantage of the inherently higher transconductance that it provides even though it requires a more complex process. The transconductance—the rate of change of drain current with respect to a change in gate voltage—is directly proportional to the charge carrier mobility in the surface inversion layer. In the case of an n-channel device, the carrier is the electron, while in a p-channel device, the charge carrier is the hole. Equation (5) is an expression for drain current of the CHEMFET. It is derived from Eq. (1) by inserting the appropriate electrochemical definitions into the expression for threshold voltage. In Eq. (5) the dependence on charge carrier mobility is explicit.

The electron mobility is more than double the hole mobility. The transconductance ratio of two otherwise geometrically identical devices that differ in type will be the ratio of the mobilities. This advantage is not without a price, however. Fabrication of n-channel devices is most conveniently done with the use of an ion implantation machine, whereas the p-channel device is made without one. The equation for the threshold voltage of a CHEMFET, Eq. (7), contains the term Q_{ss}. This represents a layer of positive electric charge that always appears near the SiO_2–Si interface (Deal et al., 1967). This charge is always positive, regardless of the conductivity type of the silicon. The practical effect is that the surface of p-type silicon under a thermally grown oxide is depleted of holes and may even be inverted to n-type. On the other hand, the surface of oxidized n-type silicon is made more n-type. Unless steps are taken to correct it, this surface inversion layer provides a parasitic current leakage path in n-channel structures. By far the most successful method for correcting this problem is to raise the surface doping concentration of the p-

type substrate by implanting acceptor impurities. This must be done selectively, avoiding the areas that become channel regions.

Following is a description of the essential steps in the design and fabrication of this particular CHEMFET.

1. N-Channel Metal Gate Process

The devices described here are implemented in n-channel, aluminum metal technology. This process uses five separate photo masking steps, each of which requires a separate and distinct pattern. Other technologies will differ substantially in the details of implementation. Another example, that of the silicon gate, is described later.

2. Masks

The first step is the creation of a composite drawing (Fig. 5). This is an accurate scale drawing that depicts each of the five layers and exactly defines the devices to be built. The drawings are normally made with the linear dimensions several hundred times actual size. Each layer is depicted by a different symbol. It is, however, more instructive to display each layer separately. This drawing may be made by hand with conventional drafting equipment, but it is more likely to be made by using special, computerized, automated design equipment developed for the microelectronics industry (Fig. 6).

Figure 7 defines the areas of the chip that are converted to n-type silicon by diffusion. It also defines the border of each individual chip. At the end of the process, no oxide or nitride will remain in these border areas. This allows separation of the individual devices. When finished, the conventional metal gates will be located in areas (1) and the chemically sensitive gates will be located in areas (2). The length of the channel, L, is defined by the spacing between the n-type diffusions at locations (2). In the devices described here, this dimension is 10 μm. The length and width of the diffused regions are determined by such considerations as allowable series electrical resistance and desired geometric shape of the device.

Figure 8 shows the second layer, which in this particular process defines an ion implant mask. When properly aligned with the previous pattern, this mask will leave photoresist covering the gate areas of both the chemically sensitive and conventional devices. This layer of photoresist will be thick enough to block boron ions to be implanted in the surface of the silicon in all nongate areas (the field areas). As previously described, these implanted ions prevent the formation of an n-type inversion layer at the surface of the silicon. The implant must be kept out of the gate areas if the threshold voltage is to be set to the desired value.

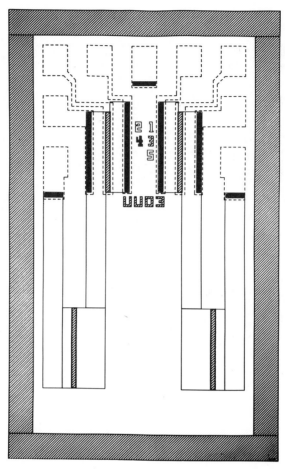

Fig. 5. Composite drawing of dual-gate CHEMFET. This computer-generated drawing shows each of the five separate layers.

Figure 9 shows the pattern of the next mask. This mask carries the definition of both gate areas and the locations of the electrical contacts between the metal and the diffused layers. Both of these are needed on this mask because of the particular sequence of processing steps. Following the diffusion of the donor impurities, a relatively thick layer of SiO_2 is grown over the entire surface of the silicon. This oxide is approximately 10 times the thickness needed for the gate insulator. The most repeatable way to achieve the oxide thickness needed for the gates is to remove all of

Fig. 6. Computer-aided design system. (Photograph courtesy of Computervision Corporation, Bedford, Massachusetts.)

it in the gate regions and then regrow it to the proper thickness. If the contacts are opened at this time, they will also be covered by the oxide layer. However, the chemical etching that reopens the contacts following the next photo step will be accomplished in a very short time, so that even if there is a defect in the mask, not all of the oxide in the defect areas will be etched away. This improves the tolerance of the process to such masking defects. This defect tolerance might not be necessary if only single CHEMFETs were ever produced by this process, but will be of significant help if the process is used to produce larger chemically sensitive integrated circuits. The field regions of the device need to be covered with an oxide layer much thicker than that at the gate regions to eliminate undesirable parasitic transistors. In principle, a parasitic transistor will be formed wherever a metal line crosses two diffusions. The only difference between the parasitic and actual transistors is a qualitative one. If the field oxide is very thick, the threshold voltage of the parasitic transistors in the field area will be higher than any applied voltage.

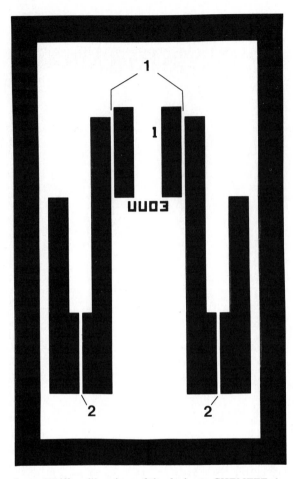

Fig. 7. First layer. "Diffused" regions of the dual-gate CHEMFET shown in Fig. 5.

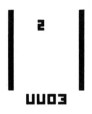

Fig. 8. Ion implant mask.

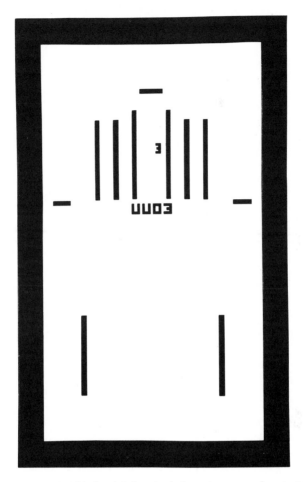

Fig. 9. Third mask. This level defines both the gate areas and contact holes.

Figure 10 is the mask pattern that is used to reopen the contact holes after forming the gate insulator. This mask has on it only the contact openings where the metal pattern makes ohmic contact with the silicon.

Figure 11 is the metal pattern. Following the photo step using this mask, the metal will be left on the device in areas defined by this pattern. During all of the other photo and etching steps in this process, material is removed in the areas defined by the mask. The only change required for this is to change the photoresist type from positive to negative, or vice versa, depending on the combination used. Or the mask could be changed

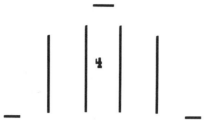

Fig. 10. Contact holes. After formation of the gate insulator it is necessary to reopen the contact holes.

from "clear field" to "dark field," depending on the field used for the other photo masking steps. In any case, this kind of process detail can only be specified in the context of the particular processing facilities to be used. After the metal has been deposited and patterned, the devices are complete.

The areas numbered 1 through 9 in Fig. 11 are the bonding pads—the points at which wires are bonded to the chip. On the final device these pads are 125 μm^2. Pads 1 and 9 are the connections to the drains of the two CHEMFETs. Pads 2 and 8 are the connections to the sources of the two CHEMFETs and simultaneously form connections to the two drains of the metal gate devices. These pads are shared to reduce the total number of connection points. Pads 4 and 6 are source connections to the metal gate devices and pads 3 and 7 are the metal gates. Pad 5 is a connection to the silicon substrate. The circuit diagram of this chip is shown in Fig. 12.

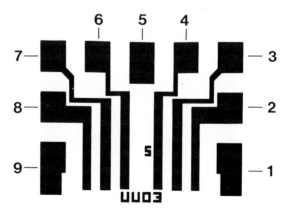

Fig. 11. Metal pattern. Bonding pads are numbered 1–9 corresponding to numbered points in Fig. 12.

In most processes, one additional layer is used. After completion of the devices through the metal layer, at which time they are complete, a layer of either silicon dioxide or silicon nitride is formed over the entire surface by some suitable chemical vapor deposition process. Silicon dioxide can be deposited by a straightforward process in which a dilute silane and oxygen mixture is passed over the surface at about 425°C. However, at this stage in the process, silicon nitride can only be deposited by a plasma-assisted method. Once aluminum is on the device, the temperature to which it can be exposed is limited. The silicon–aluminum eutectic temperature of 573°C sets an absolute upper limit, but practical considerations such as the solubility of silicon in aluminum limit the maximum temperature to which the devices may be exposed to less than 500°C. Silicon nitride is deposited at such low temperatures with the use of a low-pressure plasma. This layer of insulator above the metal on the device serves two purposes. It provides much needed mechanical protection against both scratches and electrical shorts due to dust particles, which often provide a current path between the closely spaced metal lines (10 μm in this device). For the CHEMFET devices, this layer can also be used as additional protection against the chemical solutions to which the device will be exposed. For use with this CHEMFET design, this protec-

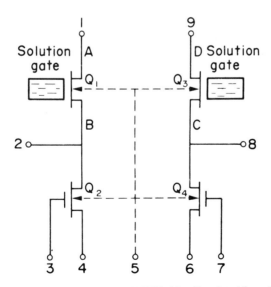

Fig. 12. Circuit schematic of dual CHEMFET chip. (Reprinted from Janata and Huber, 1980.)

tive layer is normally removed only over the chemically sensitive gate areas and the metal bonding pads.

C. MASK MAKING—HIGH-RESOLUTION REQUIREMENTS

Masks for use in the process must be produced on highly accurate, specialized equipment. As the device consists of a layered structure, the patterns on each successive layer must have the proper spatial relationship to each other. The contact opening to allow a metal contact to a diffused layer must fit entirely within the diffusion pattern, and so on. The chip example used in this chapter has minimum feature sizes of 10 μm, which in the context of today's microelectronics capability is rather large. Nevertheless, it requires a dimensional tolerance of the order of 1 μm or less. Several discussions of mask-making tolerances may be found in the literature (Bracken and Rizvi, 1983; Sze, 1983).

The drawings that define the device are usually converted to a set of high-contrast photographs, one for each layer, which are 10 times the final device size. These photographs are always on a glass plate for dimensional stability. Either a high-resolution photographic emulsion or chrome on glass can be used. The final mask for use in fabrication is made on a special photoreduction system that reduces the 10× plates to the final size and at the same time forms an array of identical patterns on another photo plate. The positional tolerances of the individual patterns are as critical as the internal tolerances on each pattern. The resulting set of multiple array plates is used in the fabrication process.

D. PROCESSING STEPS

The actual "recipe" for fabrication of the CHEMFETs will depend on the production facility. In any case, it will consist of a particular sequence of steps, previously described, known as a "run sheet." A run sheet for the CHEMFETs produced in the author's laboratory is given in the Appendix.

The starting material is p-type silicon with the $\langle 100 \rangle$ crystallographic surface as the wafer surface. The Q_{ss} charge density is lower on this face, an advantage in the n-channel device. The wafer is first chemically cleaned and then oxidized. The oxide layer is a few hundred nanometers thick. The first of the photosteps that follows opens holes in this oxide layer through which the donor impurities will be diffused. Next the wafer undergoes a standard phosphorus diffusion, during which the silicon exposed by the first photo step is converted from p-type to n-type. The next step is another oxidation step in which the oxide thickness in the field areas is increased and the diffused regions are covered with about 400 nm of SiO_2. During this step the donor atoms diffuse somewhat deeper so that the final junction is close to 1.5 μm. Next is the second photo step, in which all of the gate areas are left covered with about 1 μm of photoresist. This layer of resist serves as a mask to prevent ions implanted in the following step from reaching the silicon surface. Next follows a boron ion implantation. The ion energy is chosen so that the penetration will be slightly more than the field oxide thickness. The added thickness of the photoresist over the gate areas is enough to protect the silicon in those areas from being implanted. This step will prevent the surface of the silicon in the inactive areas from inverting to n-type because of minority carriers attracted by the positive charge Q_{ss} in the oxide.

Following removal of the photoresist, the wafers are given a short high-temperature step to anneal whatever radiation damage may have been caused by the ion implantation. Next, the third photo step is carried out. During this step the oxide is etched away in those areas which will become either gates or contact windows where the subsequent aluminum metal will make ohmic contact with the silicon.

Following this photo step, the wafers are again carefully cleaned in peroxide solutions and then oxidized to form the 80 mm of gate oxide. Then the entire wafer is coated with a layer of silicon nitride about 80 mm thick. The gate structure is now a bilayer consisting of thermally grown silicon dioxide next to the silicon capped by an unbroken layer of silicon nitride.

After nitride deposition, the wafers are coated by chemical vapor deposition with a thin layer of SiO_2. Because of equipment limitations in our laboratory we etch silicon nitride with hot phosphoric acid and the deposited SiO_2 is used as an etching mask. The fourth photo step, which follows, reopens the contact holes that were blocked in the preceding steps with the oxide and nitride layers.

In the next step, the wafers are coated with aluminum by electron beam evaporation. As previously described, an alloy of aluminum and 1% silicon is preferred but, again because of equipment limitations in our laboratory, we are using pure aluminum. The fifth photo step patterns the aluminum. As a final step, the wafers are given a low-temperature anneal in an atmosphere containing hydrogen. This step forms the ohmic contacts and reduces the effective density of interface states at the silicon–silicon dioxide interface. At this point, the wafers are ready for electrical testing.

As previously discussed, the finished devices are coated with a protective layer, usually SiO_2, deposited by low-temperature chemical vapor deposition. A final photo step opens holes through this final layer in the chemically sensitive gate areas and over the bonding pads. We use several different patterns, depending on the specific application.

E. COMPUTER AIDS

As part of the growing complexity and sophistication of integrated circuit technology, a very extensive set of computer aids for all steps of the process has been developed. It is possible to simulate most devices, processes, and circuits in great detail before actually undertaking fabrication. These computer aids can be divided into several different groups: device simulators, process simulators, time domain circuit simulators, and logic simulators. The first two of these categories are very useful in CHEMFET fabrication. The time domain simulators will become so as larger integrated circuits incorporating chemical sensors are developed. Use of these computer techniques is now standard procedure in most semiconductor fabrication operations.

1. Device Simulators

As the physical dimensions of the devices have decreased, the equations that have traditionally been used, Eqs. (1–5), lose validity because some of the basic assumptions are no longer true. These equations are one-dimensional in character, but in the very small devices the depth of the pn junctions is comparable to the lateral dimensions. In this case, the device must be treated as a two-dimensional structure. Large computer programs are available that solve numerically in two-dimensions the coupled Poisson equation and the current continuity equation (Mock, 1973). At present, the dimensions of the chemically sensitive devices used are large enough that this approach is not needed.

2. Process Simulators

The steps used in the planar fabrication process are well characterized and in many cases understood well enough that excellent mathematical models of each step are available. The kinetics of the thermal oxidation of silicon are well known (Deal and Grove, 1965) and solid state diffusion processes are now understood well enough that they can be accurately modeled mathematically. SUPREM (Ho et al., 1983) is a numerical process simulation program that is in very wide use throughout the semiconductor industry. The input to the program is a sequence of statements that give the conditions for each step of the process. An oxidation step is characterized by time, temperature, oxidizing atmosphere, and so on.

Diffusion steps and other steps are described by similar sets of quantities. The program then calculates the resulting semiconductor structure. Oxide thicknesses, dopant concentration profiles, junction depths, and some electrical properties are given. The threshold voltage for the given structure is calculated. This program is very useful in the design of a process for a CHEMFET.

3. Circuit Simulator

There are numerous computer programs that calculate the currents and voltages internal to the circuit as a function of time, although use of such programs is not necessary when dealing with a simple CHEMFET. In this case, the simple equations and experimental measurements serve the experimenter well. However, when designing an integrated circuit with chemical sensors, use of computer simulation may be an economic necessity. This is because the operation of a circuit cannot be tested until the entire fabrication process is completed; if a design error is found at this time, the entire process must be repeated at almost the same cost. One of the most widely used programs for simulation of integrated circuits is SPICE-2 (Nagel, 1975).

III. Some Particulars of CHEMFET Design and Fabrication

The design of CHEMFETs presents a number of challenges. All of the structure except the chemically sensitive part must be protected from the generally harsh chemical environment. Conventional integrated circuits can be totally encapsulated in materials carefully chosen to prevent any penetration of the chemical environment. Obviously, this is not possible with CHEMFETs.

Generally some attempt is made to separate the chemically sensitive gate from the rest of the chip, particularly the bonding pads, where electrical contact is made to the chip. In the design described in the previous section, the chemically sensitive gates are located near one end of the chip while the bonding pads are located toward the other end (Fig. 5). Such an arrangement greatly aids proper encapsulation of the device.

One method of aiding encapsulation that has been quite successful in the author's laboratory is to cover the entire top of the chip with a layer of silicon nitride deposited at high temperature. (See the fabrication procedure in the Appendix.) This layer is deposited just prior to the opening of the contact holes. In fact, the contact holes are the only openings in the nitride layer, except, of course, the sides and back of the chip. Some workers have even built ion-sensitive field effect devices in which the entire device, including the edges and back, is covered with silicon nitride (Matsuo and Esashi, 1981).

Mechanical integrity of this nitride is crucial to the stability of the device. The existence of small spots of electrically weak nitride can have a devastating effect on the stability of devices immersed in solution. A very sensitive test that is useful in evaluating the suitability of a particular nitride for CHEMFETs is to measure the ability of a thin film of the nitride to prevent the passage of current between a silicon wafer coated with the nitride and an ionic solution (Cohen *et al.*, 1978). Depending on the actual deposition conditions, nitrides vary greatly in their ability to withstand this test.

When using a nitride–oxide layer in the manner described here, the gate insulator of all of the field effect transistors, the conventional MOS-FETs as well as the CHEMFETs, will be the layered structure of silicon dioxide and silicon nitride. There are problems of electrical stability with this structure. Both the nitride layer and the interface between the nitride and oxide contain a large density of charge traps. If the electric field at the interface becomes strong enough to cause charge injection, or if nearby *pn* junctions are avalanched and generate hot electrons (Cottrell *et al.*, 1979) with sufficient energy to enter the oxide, a significant number of charges will become trapped in the insulator. The result of this space charge is a change in the flat-band voltage, hence a change in the threshold voltage. The ease of this charge injection varies with the oxide thickness, becoming less as the oxide becomes thicker. In fact, one form of semiconductor memory circuit—the electrically alterable MNOS read-only memory (EAROM)—is based on this effect. To reduce this effect the oxide layer must be kept several hundred angstrom units thick. However, even with this thick oxide layer, unpredictable threshold shifts can be induced by static charge placed on the insulator surface during handling of the unprotected devices. Antistatic precautions are required during all stages of handling of the devices.

Another structural feature that sometimes aids in encapsulation is "junction isolation" (Harame *et al.*, 1981). In this method, an island of *p*-type silicon provides the substrate for the *n*-channel CHEMFETs. This island is completely surrounded by *n*-type silicon. A reverse bias voltage on the junction electrically isolates the active CHEMFET devices. A cross section is shown in Fig. 13. With this arrangement, the encapsulation must extend beyond the isolation junction, but need not cover the sides or back of the chip. This sometimes greatly lessens the problem of encapsulation because any leakage current to the *p*-type silicon will not affect the CHEMFET's performance.

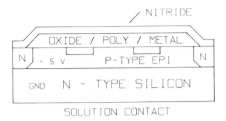

Fig. 13. Cross section of junction-isolated CHEMFET. The device is built in the island of *p*-type epitaxial silicon. "Poly" refers to polycrystalline silicon, sometimes used as a gate electrode.

The sensitivity to static charge previously discussed has led to the development of an electrostatically protected CHEMFET structure (Smith, 1982). In this design, a gate electrode is added to the CHEMFET along with an on-chip control switch (Fig. 14). The gate electrode (a), formed of polycrystalline silicon, is buried under the silicon nitride layer. In the ideal case, the impedance of the control switch (b) is infinite when the switch is open, so that the gate electrode can be electrically isolated.

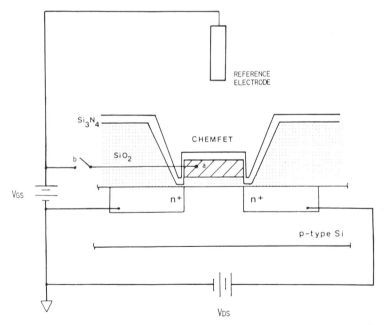

Fig. 14. Electrostatically protected CHEMFET. (Reprinted from Smith, 1982.)

This enables the gate to follow the potential developed at the solution–insulator interface in exactly the same manner as in the conventional CHEMFET. However, the impedance of the control switch is never infinite. The on-chip control switch is a normal enhancement mode MOSFET. When turned "off" it has the highly nonlinear impedance of a *pn* junction with a reverse bias. Such a device can be used to sense the changes in the electrochemical potential, but will not maintain a steady-state response.

A second type of electrostatically protected CHEMFET was built with the gate electrode exposed to the solution via a platinum contact (Smith, 1982). A cross section of this structure is shown in Fig. 15. The chemically sensitive membrane is placed on the Pt contact. If the solution–membrane–platinum structure can supply the charge faster than it leaks off through the *pn* junction to which the gate is connected, the gate potential will be near the solution–membrane electrochemical potential. When the on-chip control switch is closed, the CHEMFET gate electrode is connected to an external voltage source. This allows the CHEMFET to be electrically characterized independently of the electrochemical structure, something that is sorely lacking in the conventional device.

Another problem that has been observed with simple CHEMFETs is poor adhesion of the ion-sensitive membrane. In order to achieve maximum selectivity, the composition of the membrane must be carefully chosen. But a membrane of the proper composition for the electrochemical response generally does not adhere well to the CHEMFET gate insula-

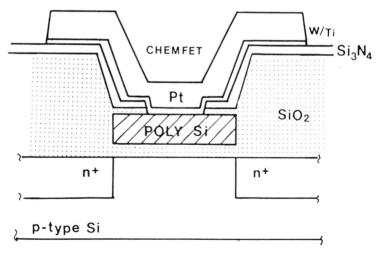

Fig. 15. Cross section of CHEMFET with platinum gate contact. (Reprinted from Smith, 1982.)

tor. If the composition is changed to give better mechanical properties, the electrical response is compromised. Poor adhesion results in gradual detachment of the membrane from the gate, allowing an electrical shunt around the membrane. Under this condition, any measurement of the electrical potential is in error. One solution to this problem has been the development of a three-dimensional suspended mesh above the chemically sensitive gate (Blackburn and Janata, 1982). This structure serves to anchor the membrane in place and greatly extends the useful life of the device. The suspended mesh is shown in Fig. 16.

IV. Encapsulation

Encapsulation of CHEMFETs is one of the most important steps in the whole fabrication sequence. The very nature of CHEMFETs requires that they be used in solutions of electrolytes, which, from the electronic point of view, are hostile environments. Electrical integrity of the whole CHEMFET assembly is a critical prerequisite for successful operation of these devices.

There is no general recipe for encapsulation. The choice of materials is governed by the general device package (catheter, transistor header, etc.), by the geometry and layout of the chip, and by the intended use. From the point of view of operation of the device, there are two distinct encapsulation problems:

1. The encapsulation and electrical integrity of the gate and its immediate surrounding. Electrical leakage in this area will usually have an adverse effect on the electrochemical behavior of the transistor and must be prevented.

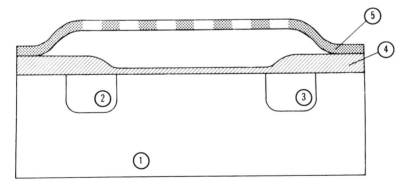

Fig. 16. Suspended mesh CHEMFET. (1) Substrate; (2 and 3) source and drain; (4) insulator. The ion-sensitive membrane is held in contact with the gate insulator by the suspended mesh (5). (From Blackburn and Janata, 1982. Reprinted by permission of the publisher, The Electrochemical Society, Inc.)

2. The encapsulation and electrical integrity of the rest of the package. This affects the lifetime of the probe and can be an important consideration in electrical safety in medical applications.

The choice of materials has to be based on the following criteria: (1) low bulk permeability for water and electrolytes, (2) good adhesion between materials used, and (3) the mechanical and rheological properties of the noncured encapsulant, suitable for the fabrication.

The most satisfactory combination of materials we have used has been found by trial and error over several years. Silicon rubbers and epoxy resins are generally recognized as good encapsulants for electronic components. Although the uptake of water by silicone rubber is low, its permeability is considerably higher than that of epoxy resins. Another reason for preference of epoxy as the encapsulant is that silicone rubber does not adhere well to PVC (membrane and catheter). The adhesion of epoxy to the silicon chip coated with silicon nitride can be improved by silanization. This also improves the adhesion of the PVC membrane (see below) to the gate area of the chip.

The optimum encapsulant used for hand encapsulation is a high-grade epoxy resin, such as Epon 825, with a suitable diamine as the cross-linker. We have used Jeffamine D230, which gives a relatively hard, dried cross-linked resin with good adhesion to both silicon nitride and PVC membranes. One of the most difficult steps during encapsulation by hand is the definition of the gate areas. For this step, we are using a mixture of Epon 825 (73% w/w) and Jeffamine D230 (27% w/w). To this we add 12% w/w fumed silica (Silanox 101) for thixotropicity. The encapsulation around bonding wires is done with 5% Silanox epoxy. After the initial application, these materials are allowed to cure at room temperature for 12 h and then at 60°C for 12 h. A photograph of a hand-encapsulated ISFET chip is shown in Fig. 17.

The electrical integrity of the gate is critically important for predictable operation of the ISFETs with membranes (i.e., other than bare SiO_2 or Si_3N_4 gate). Any passage of current through the gate "insulator" and the overlying membrane causes polarization of the latter. This leads to instability of the potential and, after prolonged polarization, sometimes to degradation of the membrane. This situation is equivalent to using a low-input-impedance amplifier to measure voltage across a high-resistance ISE membrane.

We have undertaken a detailed study (Cohen et al., 1978) of the electrical integrity of SiO_2 and Si_3N_4 prepared by chemical vapor deposition and RF sputtering and have found that Si_3N_4 prepared by either technique

Fig. 17. Photomicrograph of hand-encapsulated dual-gate CHEMFET.

is an excellent insulator. These materials can withstand exposure to 0.15 M NaCl for 90 days. On the other hand, thermally grown SiO_2 lost its insulating property within hours after immersion in the solution. The mechanism of this breakdown appears to be consistent with the formation of solution-filled microfissures in the top layer of the SiO_2, and not with total hydration of the SiO_2 layer, as originally thought (Bergveld, 1972; Moss et al., 1975; Janata and Moss, 1976). When a sufficiently high voltage is applied, the electric field at the tip of the fissure is much greater than that on the flat surface of the insulator. This leads to catastrophic breakdown of the remaining insulator beneath the fissure. Depending on the polarity of the applied voltage, one or the other of the following reactions then takes place:

$$\text{Silicon negative:} \quad H_2O + e \longrightarrow \tfrac{1}{2} H_2 + OH^- \tag{I}$$

$$\text{Silicon positive:} \quad Si + 2 H_2O \longrightarrow SiO_2 + 4 H^+ + 4 e \tag{II}$$

The first reaction leads to the evolution of hydrogen bubbles, which can be observed (Bergveld, 1972; Moss et al., 1975; Janata and Moss, 1976) under a microscope, thus aiding in location of the leak. The presence of microcracks in SiO_2 has been suspected by Esashi and Matsuo (Esashi and Matsuo, 1975a,b) and by Schenk (1978).

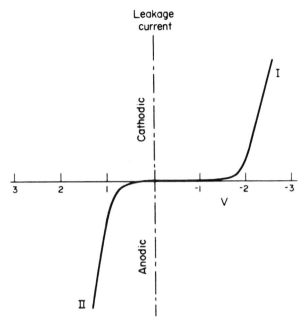

Fig. 18. Current leakage from a poorly encapsulated CHEMFET. (Reprinted from Janata and Huber, 1980.)

The ultimate test of the electrical integrity of the whole package is to connect all wires of the ISFET (i.e., drain, source, and substrate) together and to polarize the device from +3 to −3 V against a low-resistance ($R <$ 100 kilohms) reference electrode. The shape of the current–voltage curve (Fig. 18), as well as the absolute value of the leakage current, reveals the possible leak. Increase of leakage current on the cathode side corresponds to evolution of hydrogen [reaction (I)]. The value of the decomposition potential depends on the material that forms the substrate for this reaction (a wire or Si) and also on the previous polarization history. The anodic branch of the leakage current is due to reaction (II) if the leak is through Si. On the metal, the corresponding reaction can be the formation of oxide or insoluble salt or, eventually, evolution of oxygen. Since all these reactions are pH-dependent, it follows that the leak does not make a good reference electrode if the pH of the solution is changing. This can account for inconsistent data obtained with the SiO_2 gate pH ISFETs. It is also apparent from Fig. 18 that testing between +1 and −1 V cannot tell anything about the encapsulation, a point that is not generally appreciated.

Appendix: *Metal Gate CHEMFET Fabrication Run Sheet*

The actual recipe for fabrication of the CHEMFETs will depend on the production facility. Following is a detailed outline of the steps that have produced good devices in the author's laboratory.

Because of the very stringent cleanliness requirements, the wafers must be carefully cleaned before any high-temperature processing is carried out. We use a two-step process designed to remove residual amounts of both organic material and metals. In the first step we use a mixture of NH_4OH and H_2O_2, and in the second step a mixture of HCl and NH_4OH. This cleaning procedure works very well in a laboratory setting where only small numbers of wafers are processed together.

A. SILICON MATERIAL

> p-type, boron doped
> 5 to 10 ohm-cm resistivity
> $\langle 100 \rangle$ orientation
> polished for MOS use

B. INITIAL CLEAN

> 10 min at 80°C in 5 : 1 : 1 mixture of $H_2O : NH_4OH : H_2O_2$
> deionized water rinse
> 10 min at 80°C in 5 : 1 : 1 mixture of $H_2O : HCl : H_2O_2$
> deionized water rinse
> 15 s in 10 : 1 mixture of $H_2O : HF$
> spin dry
> grow 500 nm of oxide at 950°C
> 15 min dry O_2
> 95 min wet O_2
> 15 min dry O_2

C. DRAIN–SOURCE PHOTO

> spin coat HMDS
> spin coat negative photoresist
> prebake 90°C, 30 min
> expose first mask
> develop
> inspect under microscope
> postbake 120°C, 30 min
> etch—buffered HF 5 min to dewet
> deionized water rinse
> dry

D. PHOTORESIST STRIP

> strip in the appropriate solution for the resist used
> deionized water rinse
> dry

E. SOURCE–DRAIN DIFFUSION

all steps carried out at 950°C
preheat in nitrogen 3 min, 950°C
oxidation (N_2 plus O_2) 10 min, 950°C
phosphorus source, $POCl_3$, on for 23 min
purge with N_2, 5 min

F. PHOSPHORUS GLASS REMOVAL

buffered HF etch—15 s
deionized water rinse

G. FIELD OXIDATION/DRIVE-IN

temperature 1000°C
dry O_2, 10 min
wet O_2, 60 min
dry O_2, 10 min

H. FIELD ION IMPLANT PHOTO

spin coat HMDS
spin coat negative photoresist
prebake 90°C, 30 min
expose mask number 2
develop
inspect under microscope
DO NOT PREBAKE

I. ION IMPLANT

boron ion implant
beam current 1.5×10^{-6} A
ion energy 160 keV
Dose 3.0×10^{12} ions/cm^2

J. PHOTORESIST STRIP

strip in the appropriate solution for the resist used
deionized water rinse
dry

K. WAFER CLEAN

10 min at 80°C in 5:1:1 mixture of $H_2O : NH_4OH : H_2O_2$
deionized water rinse
10 min at 80°C in 5:1:1 mixture of $H_2O : HCl : H_2O_2$
deionized water rinse
15 s in 10:1 mixture of $H_2O : HF$
spin dry

L. ION IMPLANT ANNEAL

temperature 950°C
dry N_2, 5 min
dry O_2, 5 min
dry N_2, 5 min

M. PHOTO STEP—DEFINE GATE AND CONTACT AREAS

spin coat HMDS
spin coat negative photoresist
prebake 90°C, 30 min
expose mask number 3
develop
inspect under microscope
postbake 120°C, 30 min
etch—buffered HF 5 min to dewet
deionized water rinse
dry
inspect

N. PHOTORESIST STRIP

strip in appropriate chemical strip

O. WAFER CLEAN

10 min at 80°C in 5 : 1 : 1 mixture of $H_2O : NH_4OH : H_2O_2$
deionized water rinse
10 min at 80°C in 5 : 1 : 1 mixture of $H_2O : HCl : H_2O_2$
deionized water rinse
15 s in 10 : 1 mixture of $H_2O : HF$
deionized water rinse
spin dry

P. GATE OXIDATION CYCLE

grow 80-nm SiO_2 in dry oxygen
growth time approximate 120 min

Q. DEPOSIT SILICON NITRIDE

deposit 80 nm Si_3N_4
temperature 1000°C
chemical vapor deposition (CVD) using $SiCl_4$ and NH_3
atmospheric pressure

R. DEPOSIT SiO_2

CVD deposition of 125 nm SiO_2
425°C, atmospheric pressure
densify at 950°C, 10 min in dry O_2

S.   Photo Step—Open Contact Holes

 spin coat HMDS
 spin coat negative photoresist
 prebake 90°C, 30 min
 expose mask number 4
 develop
 inspect under microscope
 postbake 120°C, 30 min
 etch—buffered HF, 80 s
 deionized water rinse
 dry
 inspect

T.   Photoresist Strip

 strip in appropriate chemical strip

U.   Silicon Nitride Etch

 phosphoric acid at 180°C, 20 min
 boiling deionized water rinse, 5 min
 cold deionized water rinse, 5 min
 etch dilute HF $(1 : 10 \ H_2O : HF)$
 10 deionized water rinse (dry)

V.   Aluminum Deposition

 deposit 100 nm aluminum
 electron beam vacuum evaporation

W.   Aluminum Photo Step

 spin coat HMDS
 spin coat with positive resist
 prebake 90°C, 20 min
 expose mask number 5
 develop
 inspect
 postbake 120°C

X.   Aluminum Etch

 etch in phosphoric—nitric—acetic acid mixture
 53–55°C for 90 s
 deionized water rinse
 dry

Y.   Photoresist Strip

 strip in appropriate chemical strip

Z. ALUMINUM ANNEAL

anneal at 450°C
dry N_2, 10 min
dry N_2 + H_2, 20 min
dry N_2, 10 min

AA. ELECTRICAL TEST

REFERENCES

Adams, A. C. (1983). *In* "VLSI Technology" (S. M. Sze, ed.), Chapter 2, pp. 93–128. McGraw-Hill, New York.
Bergveld, P. (1972). *IEEE Trans. Biomed. Eng.* **BME-19**(5), 342–351.
Blackburn, G., and Janata, J. (1982). *J. Electrochem. Soc.* **129**, 2580–2584.
Bracken, R. C., and Rizvi, S. A. (1983). *In* "Microlithography in Semiconductor Device Processing" (N. G. Einspruch and G. B. Larrabee, eds.), Vol. 6, Chapter 5. Academic Press, New York.
Chapman, B. (1980). "Glow Discharge Processes." Wiley, New York.
Cohen, R. M., Huber, R. J., Janata, J., Ure, R. W., and Moss, S. D. (1978). *Thin Solid Films* **53**, 169–173.
Collins, R. H., and Deverse, F. T. (1970). U.S. Patent 3,549,368.
Cottrell, P. E., Troutman, R. R., and Ning, T. H. (1979). *IEEE Trans. Electron Devices* **ED-26**(4), 520–533.
Deal, B. E., and Grove, A. S. (1965). *J. Appl. Phys.* **36**, 3770.
Deal, B. E., Sklar, M., Grove, A. S., and Snow, E. H. (1967). *J. Electrochem. Soc.* **114**, 266.
Elliott, D. J. (1982). "Integrated Circuit Fabrication Technology." McGraw-Hill, New York.
Esashi, M., and Matsuo, T. (1975a). Proceedings of the 6th Conference on Solid State Devices, Tokyo, 1974. *J. Jpn. Soc. Appl. Phys. Suppl.* **44**, 339–343.
Esashi, M., and Matsuo, T. (1975b). *IEEE Trans. Biomed. Eng.* **BME-25**, 184.
Grove, A. S. (1967). "Physics and Technology of Semiconductor Device." Wiley, New York.
Harame, D., Shott, Plummer, J., and Meindl, J. (1981). *Tech. Dig.—Int. Electron Devices Meet.*, pp. 467–468.
Ho, C. P., Plummer, J. D., Hansen, S. E., and Dutton, R. W. (1983). *IEEE Trans. Electron Devices* **ED-30**(11) 1438–1453.
Hoffman, V. (1978). *Solid State Technol.* **21**(12), 47–56.
Janata, J., and Moss, S. D. (1976). *Biomed. Eng.* **11**, 241.
Janata, J., and Huber, R. J. (1980). *In* "Ion-Selective Electrodes in Analytical Chemistry" (H. Freiser, ed.), Chapter 3, pp. 107–174. Plenum, New York.
Katz, L. E. (1983). *In* "VLSI Technology" (S. M. Sze, ed.), Chapter 4, pp. 131–167. McGraw-Hill, New York.
Kern, W., and Puotinen, D. A. (1970). *RCA Rev.*, pp. 187–206.
Kilpatrick, M. K. (1984). *Solid State Technol.* **27**(3) 151–155.
McBride, P. E., Janata, J., Comte, P. A., Moss, S. D., and Johnson, C. C. (1978). *Anal. Chim Acta* **101**, 239.
Many, A., Goldstein, Y., and Grover, N. B. (1965). "Semiconductor Surfaces." North-Holland Publ., Amsterdam.

Matsuo, T., and Esashi, M. (1981). *Sens. Actuators* **1**(1), 77–96.

Matsuo, T., and Wise, K. (1974). *IEEE Trans. Biomed. Eng.* **BME-21**, 485–487.

Mock, M. S., (1973). *Solid State Electron.* **16**, 601–609.

Moss, S. D., Janata, J., and Johnson, C. C. (1975) *Anal. Chem.* **47**, 2238.

Muller, R. S., and Kamins, T. I. (1977). "Device Electronics for Integrated Circuits." Wiley, New York.

Nagel, L. W. (1975). SPICE-2: A computer program to simulate semiconductor circuits. Ph.D. Thesis, University of California, Berkeley.

Pearce, C. W. (1983). "VLSI Technology" (S. M. Sze, ed.), pp. 9–49. McGraw-Hill, New York.

Poate, J. M., and Tisone, T. C. (1974). *Appl. Phys. Lett.* **24**(8), 391–393.

Ravi, K. V. (1981). "Imperfections and Impurities in Semiconductor Silicon." Wiley, New York.

Rohatgi, A., Butler, S. R., and Feigl, F. J. (1979). *J. Electrochem. Soc.* **126**, 149.

Schenk, J. F. (1978). "Workshop on Theory, Design, and Biomedical Applications of Solid State Chemical Devices." CRC Press, Cleveland, Ohio.

Shockley, W. (1952). *Proc. IRE* **40**, 641.

Seidel, T. E. (1983). *In* "VLSI Technology" (S. M. Sze, ed.), Chapter 6, p. 219–264. McGraw-Hill, New York.

Smith, R. L. (1982). Ion-sensitive field effect transistors with polysilicon gates. Ph.D. Thesis, University of Utah, Salt Lake City.

Sze, S. M. (1981). "Physics of Semiconductor Devices." Wiley, New York.

Sze, S. M., ed. (1983). "VLSI Technology." McGraw-Hill, New York.

Taubenest, R., and Ubersax, H. (1980). *Solid State Technol.* **23**(6), 74–79.

Van Gelder, W., and Hauser, V. E. (1967). *J. Electrochem. Soc.* **114**, 869.

Zemel, J. N. (1975). *Anal. Chem.* **47**, 255A.

4

An Introduction to Piezoelectric and Pyroelectric Chemical Sensors

JAY N. ZEMEL

CENTER FOR CHEMICAL ELECTRONICS
DEPARTMENT OF ELECTRICAL ENGINEERING
UNIVERSITY OF PENNSYLVANIA
PHILADELPHIA, PENNSYLVANIA

I. Introduction

In 1970, P. Bergveld published his first paper on the concept of an ion-sensitive field effect transistor. Although the device described had relatively poor characteristics compared to conventional ion-selective electrodes, the paper was important because of the use of the then-emerging integrated circuit technology to produce a chemically sensitive electronic device (CSED). Progress and interest in this type of information acquisition device has grown substantially since that time (Zemel and Bergveld, 1981).

The approach presented in this chapter to the study of CSEDs requires an understanding of two key aspects of the information acquisition problem. The first aspect concerns the physicochemical phenomena that allow the conversion of concentration or activity into an electronically measur-

163

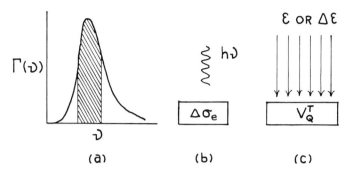

Fig. 1. Illustration of optical detection. (a) Radiation spectrum of intensity $\Gamma(\nu)$ versus frequency ν. The hatched region corresponds to the interval between ν_1 and ν_2 in the definition of the energy interval $\Delta\varepsilon$. (b) Photoconductive detector producing a change in the conductivity, $\Delta\sigma_e$, as a result of exposure to $h\nu$. (c) Photothermal detector giving rise to thermal voltage V_Q^T due to the energy ε or energy interval $\Delta\varepsilon$.

able quantity; the second concerns the methods needed to extract an electronic signal in appropriate form from a suitable device. This approach is similar in many respects to that employed for optical detection (Zemel, 1979). An optical detector responds when it is exposed to a flux of photons defined as $\Gamma(\nu)$ per unit energy $h\nu$ (Fig. 1). The information of interest is $\Gamma(\nu)$, and it must be transformed into an electrical signal by the detector. There are two photon–detector interaction processes by which this transformation can be accomplished. The first process makes use of the quantum interaction between the photons and the detector that produces free electrons and holes:

$$e + p \underset{h\nu}{\overset{\longrightarrow}{\rightleftharpoons}} 0 \tag{1a}$$

or free electrons and a trapped hole

$$e + D_+^T \underset{h\nu}{\overset{\longrightarrow}{\rightleftharpoons}} 0 \tag{1b}$$

or free holes and a trapped electron

$$p + A_-^T \underset{h\nu}{\overset{\longrightarrow}{\rightleftharpoons}} 0 \tag{1c}$$

where e and p represent the electrons and holes, respectively, D_+^T is a donor (positively charged) trap, and A_-^T is an acceptor (negatively charged) trap. The resulting change in conductivity $\Delta\sigma_e$ of the specimen can be written as (Dalven, 1980)

$$\Delta\sigma_e = \Delta\sigma_n + \Delta\sigma_p \tag{2}$$

where

$$\Delta\sigma_n = q \, \Delta n \, \mu_n + q \, \Delta\mu_n \tag{3a}$$

$$\Delta\sigma_p = q \, \Delta p \, \mu + qp \, \Delta\mu_p \tag{3b}$$

$\Delta\sigma_n$ and $\Delta\sigma_p$ are the changes in conductivity due to generation of electrons and holes, respectively; q is the electronic charge; Δn and Δp are the changes in the electron and hole densities n and p, respectively; and $\Delta\mu_n$ and $\Delta\mu_p$ are the changes in the electron and hole mobilities μ_n and μ_p, respectively.

Another mechanism for detecting optical radiation makes use of the total photon energy ε (Putley, 1970)

$$\varepsilon = \int_\nu \Gamma(\nu) h\nu \, d\nu \tag{4}$$

or the partial photon energy $\Delta\varepsilon(\Delta\nu, \bar{\nu})$

$$\Delta\varepsilon(\Delta\nu, \bar{\nu}) = \int_{\nu_1}^{\nu_2} \Gamma(\nu) h\nu \, d\nu$$

$$\Delta\nu = \nu_2 - \nu_1 \tag{5}$$

$$\bar{\nu} = \tfrac{1}{2}(\nu_1 + \nu_2)$$

By treating the incidence photon flux as a source of thermal energy, the detector undergoes an increase in temperature which changes an electronic property of the sensor, such as the thermoelectric voltage V_Q^T.

For the electromagnetic radiation to be converted into such a measurable quantity as $\Delta\sigma_e$ or V_Q^T, a variety of appropriate materials and structures can be employed. Full characterization of the optical sensor requires that the fundamental interaction processes between light and the relevant phenomena be sufficiently well understood that the limits of detectability can be defined. The situation is much the same for the CSED, the major difference being that, instead of detecting photons of different frequencies, the CSED detects atoms, ions, or molecules.

In general, very few sensors have linear responses to chemical stimuli. With the exception of avalanche detectors, most electromagnetic radiation detectors are square law devices (proportional to the square of the wave amplitude). As a result, they respond linearly to either ε or $\Delta\varepsilon(\Delta\nu, \bar{\nu})$. Electrochemical or potentiometric sensors for pH and the like respond to the logarithm of the activity rather than to the ionic activity itself. Because of this, it will be important to process information from more than one sensor in order to separate out the activities of multiple constituents in a chemical environment, even though the sensors themselves may be reasonably specific. This procedure will also require substantial computational and theoretical capability (Kowalski, 1981).

In this chapter, chemical detection is examined in the context of two specific classes of microfabricated CSEDs, each of which relies on a different electronic process for its basic sensitivity. Both structures are manufactured by the type of lithographic procedure generally employed in integrated circuit technology. The structures and the physical processes that provide their response mechanisms are:

1. The piezoelectric gravimetric sensor—mass of the chemical species.
2. The pyroelectric detector—heat of phase transformations.

Both of these devices satisfy the general requirements that have been established for microfabricated CSEDs (Zemel *et al.*, 1981).

In the piezoelectric oscillator, the natural frequency of the crystal oscillator is determined in part by the mass of the crystal. A suitably cut quartz crystal oscillator has an extremely narrow resonance band (usually referred to as a high Q) and the resonance frequency will vary with any changes in the mass of the crystal. This is the principle underlying the widely employed quartz crystal thickness monitor. A simple modification of this circuitry and the measuring crystal allows one to monitor adsorption–desorption processes on chemically sensitive layers bound to the crystal surface. Because the entire crystal oscillates to generate the resonance frequency, development of a multiple species sensor on a single wafer chip is not a trivial problem.

In the second type of device, a pyroelectric crystal responds to temperature changes in the same way as a piezoelectric crystal responds to stress, that is, with a change in the spontaneous polarization of the crystal. The change in polarization induces a charge on the electrodes; this charge can be readily measured. Because of the sensitivity to local heat flow and the limited heat flow parallel to the surface compared to the flow normal to the surface, multiple sensors on a chip appear to be more feasible with pyroelectrics than with piezoelectrics.

II. Piezoelectric and Pyroelectric Sensors

A. INTRODUCTION

There are classes of materials whose polarization vector **P** is a function of the applied stress and temperature. As the stress or temperature varies, the surface charge also varies. The resulting electrification of the surface due to a change in applied stress is called piezoelectricity, and that due to a temperature change is called pyroelectricity. In general, these phenomena occur in crystals that do not have a center of inversion symmetry. However, pyroelectricity in crystals has more severe symmetry requirements than does piezoelectricity. As an example, pyroelectricity does not arise in the widely used α-quartz single crystals. Piezo- and pyroelectricity are also observed in certain stressed polymers such as polyvinylidene fluoride (Lovinger, 1983). This polymeric material is widely used because of its low cost for large areas, relatively high piezo- and pyroelectric response, and handling. In this section, some basic properties of materials that are piezo- or pyroelectric are reviewed in terms of their utility in sensor research.

As is evident in the discussion of sensors in the other chapters of this book, there is a common tendency to assume that single-crystal silicon is the ideal material for microfabricated sensors. It is undoubtedly true that silicon is the preferred substrate, since it is simple to provide the additional on-chip signal processing without hybridization. However, as versatile as silicon is, there are physical phenomena that are suited to information acquisition (such as piezo- and pyroelectricity) that cannot occur in silicon.

Although this subject will not be discussed here, it should be noted that monolithic sensors based on deposited piezo- or pyroelectric materials on silicon may be less cost-effective for small production runs (typical of sensor devices) than hybrid structures where sensor and electronics are separately optimized. It is too early in the evolution of these devices to make hard decisions.

Piezoelectric crystals have been used for many years as stress sensors; their initial application as acoustic transducers dates back to 1917 (Mason, 1950). More recently, these devices have been used as gravimetric sensors to detect adsorption of gases (King *et al.*, 1968). The pyroelectric radiometer is widely used today in place of the thermocouple pyrometer (Putley, 1970). The electrodes on these sensors are fabricated by the same technology employed in silicon integrated circuit manufacture.

B. THE SENSOR NATURE OF PIEZOELECTRICITY

As pointed out by Mason (1950), a piezoelectric crystal is simultaneously a condenser, a motor, and a generator. Because of the high resistivity of the material, the electrical behavior is largely capacitive. However, when a stress is applied to the system, charge is generated at the plates and, according to the Onsager relations (Smith *et al.*, 1967), there

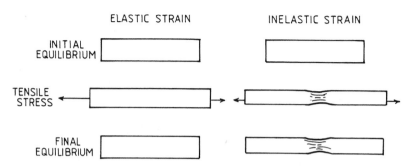

Fig. 2. The structures on the left illustrate the elastic response of a solid and those on the right illustrate an inelastic strain.

will be an inverse relation in which a charge (or voltage) on the capacitor will induce a mechanical strain.

When a solid is subjected to an external force, the shape and volume will be deformed. If the external force is small enough, the solid will return to its original shape when the force is removed. However, if the force exceeds some critical value, the solid will flow and a permanent deformation will occur. The latter regime is called the plastic flow regime, and the former is the elastic deformation regime (Fig. 2). Only the elastic region will be considered here. In this region, the stress–strain, dielectric displacement–electric field, and entropy–temperature relations are linear and the mathematics for describing sensor response is relatively straightforward. To better appreciate the response, a more detailed discussion of stress and strain is presented.

1. Stress

The force acting on a body can be decomposed into a body force $\mathbf{F_B}$, such as gravity, and surface force $\mathbf{F_S}$, such as tension. The resultant force on the body, shown in Fig. 3, will be zero at equilibrium. Thus, according to Stratton (1941),

$$\int_v \mathbf{F_B} \, dv + \int_s \mathbf{F_s} \, da = 0 \tag{6}$$

Furthermore, the moments of the forces must also be zero to avoid rotation. Therefore

$$\int_v \mathbf{F_B} \times \mathbf{r} \, dv + \int_s \mathbf{F_s} \times \mathbf{r} \, da = 0 \tag{7}$$

If $\mathbf{n}$ is the normal, outward-facing unit vector, then the component of the

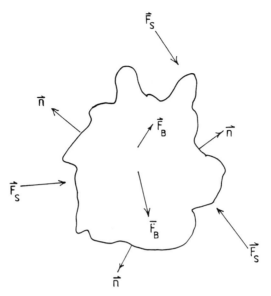

Fig. 3. The vector **n** is the outward-facing normal vector from the surface of the structure, and $\mathbf{F}_s$ are the surface forces and $\mathbf{F}_B$ the bulk forces. These forces are responsible for the strain in a solid.

surface force can be written in terms of a quantity **A** whose vector components are $\mathbf{A}_x$, $\mathbf{A}_y$, and $\mathbf{A}_z$,

$$F_{sx} = \mathbf{A}_x \cdot \mathbf{n} = A_{xx}n_x + A_{xy}n_y + A_{xz}n_z \tag{8}$$

$$F_{sy} = \mathbf{A}_y \cdot \mathbf{n} = A_{yx}n_x + A_{yy}n_y + A_{yz}n_z \tag{9}$$

$$F_{sz} = \mathbf{A}_z \cdot \mathbf{n} = A_{zx}n_x + A_{zy}n_y + A_{zz}n_z \tag{10}$$

Introducing the x component of $\mathbf{F}_B$ into Eq. (6) and combining this with the x component of $\mathbf{F}_S$ defined in Eq. (8) yields

$$\int_v F_{Bx} \, dv + \int_s \mathbf{A}_x \cdot \mathbf{n} \, da = 0 \tag{11}$$

Using the divergence theorem, it is easy to show that

$$F_{Bx} + \nabla \cdot \mathbf{A}_x = 0 \tag{12}$$

Similarly

$$F_{By} + \nabla \cdot \mathbf{A}_y = 0 \tag{13}$$

$$F_{Bz} + \nabla \cdot \mathbf{A}_z = 0 \tag{14}$$

If the x component of Eq. (7) is considered, then

$$\int_v (F_{By}z - F_{Bz}y) \, dv + \int_s (F_{sy}z - F_{sz}y) \, da = 0 \tag{15}$$

If Eqs. (12)–(14) are then used for F_{Bi}, this yields

$$\int_v F_{By}z - F_{Bz}y) \, dv + \int_s (\mathbf{A}_y z - \mathbf{A}_z y) \cdot \mathbf{n} \, da = 0 \tag{16}$$

which, when combined with the divergence theorem, yields

$$F_{By}z - F_{Bz}y + \nabla \cdot (\mathbf{A}_y z - \mathbf{A}_z y) = 0 \tag{17}$$

It is easy to show that

$$\nabla \cdot (\mathbf{A}_z y) = y \, \nabla \cdot \mathbf{A}_z + \mathbf{A}_z \cdot \nabla y = y \, \nabla \cdot \mathbf{A}_z + A_{zy} \tag{18}$$

$$\nabla \cdot (\mathbf{A}_y z) = z \, \nabla \cdot \mathbf{A}_y + \mathbf{A}_y \cdot \nabla z = z \, \nabla \cdot \mathbf{A}_y + A_{yz} \tag{19}$$

Combining Eqs. (17), (18), and (19) yields

$$(F_{By} + \nabla \cdot \mathbf{A}_y)z - (F_{Bz} + \nabla \cdot \mathbf{A}_z)y + A_{zy} - A_{yz} = 0 \tag{20}$$

Using Eqs. (13) and (14) causes the first two terms in Eq. (20) to vanish, so that

$$A_{zy} = A_{yz} \tag{21}$$

It is straightforward to demonstrate that in general

$$A_{ij} = A_{ji} \tag{22}$$

where A_{ij} is the ith component of the force acting outward in the jth direction from an element of area, and is hereafter defined as σ_{ij}. The nine components in Eq. (22) are components of the stress tensor $\mathbf{\sigma}$,

$$\mathbf{\sigma} = \begin{pmatrix} \sigma_{xx} & \sigma_{xy} & \sigma_{xz} \\ \sigma_{xy} & \sigma_{yy} & \sigma_{yz} \\ \sigma_{xz} & \sigma_{yz} & \sigma_{zz} \end{pmatrix} = \begin{pmatrix} \sigma_1 & \sigma_4 & \sigma_5 \\ \sigma_4 & \sigma_2 & \sigma_6 \\ \sigma_5 & \sigma_6 & \sigma_3 \end{pmatrix} \tag{23}$$

In tensor notation, the requirement for static equilibrium is

$$\mathbf{F}_B + \nabla : \mathbf{\sigma} = 0, \qquad \sigma_{jk} = \sigma_{kj} \tag{24}$$

2. Strain

Under the action of the body and surface forces, a solid can undergo linear and rotational displacements of the entire system as well as a relative motion or distortion of the solid. It is this distortion with reference to the center of mass of the solid that constitutes strain. If $\mathbf{r}$ is the radius vector from the center of mass of the solid to a point P in the unstressed solid and $\mathbf{r}'$ is the corresponding vector in the stressed solid, then the displacement $\mathbf{u}$ is

$$\mathbf{r} = \mathbf{r} + \mathbf{u}(\mathbf{r}) \tag{25}$$

The relative displacement of a nearby point P_1 will be

$$\mathbf{r}_1' = \mathbf{r}_1 + \mathbf{u}_1(\mathbf{r}_1) \tag{26}$$

Defining the separation between P and P_1 before deformation as

$$\delta\mathbf{r} = \mathbf{r}_1 - \mathbf{r} \tag{27}$$

then the relative separation of the two displacements $\mathbf{u}$ and $\mathbf{u}_1$ is

$$\delta\mathbf{u} = \mathbf{u}_1 - \mathbf{u} = \mathbf{u}(\mathbf{r} + \delta\mathbf{r}) - \mathbf{u}(\mathbf{r}) \tag{28}$$

Expanding $\mathbf{u}(\mathbf{r} + \delta\mathbf{r})$ about $\mathbf{r}$ yields

$$\mathbf{u}(\mathbf{r} + \delta\mathbf{r}) = \mathbf{u}(r) + (\delta\mathbf{r} \cdot \nabla)\mathbf{u}(r) + \cdots \tag{29}$$

By retaining only the linear terms

$$\delta\mathbf{u} = \delta\mathbf{r} \cdot \nabla\mathbf{u} \tag{30}$$

with component

$$\delta u_i = \frac{\partial u_i}{\partial x_j} \delta x_j = u_{ij} \delta x_j \qquad (i, j = 1, \ldots, 3) \tag{31}$$

the components u_{ij} $(i \neq j)$ can be rewritten in terms of a local shear strain and a proper rotation. In other words

$$u_{ij} = \tfrac{1}{2}(u_{ij} + u_{ji}) + \tfrac{1}{2}(u_{ij} - u_{ji}) = S_{ij} + \tfrac{1}{2}\nabla\mathbf{u}(r) \tag{32}$$

$$u_{ji} = \tfrac{1}{2}(u_{ji} + u_{ij}) - \tfrac{1}{2}(u_{ij} - u_{ji}) = S_{ij} - \tfrac{1}{2}\nabla\mathbf{u}(r) \tag{33}$$

As a result, the displacement vector can be decomposed into an expression for the elastic deformation of the solid and a proper rotation of the solid about its center of mass,

$$\delta\mathbf{u} = \mathbf{S} \cdot \delta\mathbf{r} + \mathbf{R} \cdot \delta\mathbf{r} = \delta\mathbf{u}_E + \delta\mathbf{u}_R \tag{34}$$

where $\delta\mathbf{u}_E$ is the local deformation, $\delta\mathbf{u}_R$ is the rigid rotation, and $\mathbf{R}$ is the antisymmetric rotation tensor. The symmetric second-rank tensor $\mathbf{S}$ or strain tensor has components

$$S_{11} = \frac{\partial u_x}{\partial x} = u_{xx} = S_1 \tag{35}$$

$$S_{22} = \frac{\partial u_y}{\partial y} = u_{yy} = S_2 \tag{36}$$

$$S_{33} = \frac{\partial u_z}{\partial z} = u_{zz} = S_3 \tag{37}$$

$$S_{23} = \frac{1}{2}\left(\frac{\partial u_z}{\partial y} + \frac{\partial u_y}{\partial z}\right) = \frac{1}{2}(u_{zy} + u_{yz}) = \frac{1}{2}S_4 \tag{38}$$

$$S_{13} = \frac{1}{2}\left(\frac{\partial u_z}{\partial x} + \frac{\partial u_x}{\partial z}\right) = \frac{1}{2}(u_{xz} + u_{zx}) = \frac{1}{2}S_5 \tag{39}$$

$$S_{12} = \frac{1}{2}\left(\frac{\partial u_x}{\partial y} + \frac{\partial u_y}{\partial x}\right) = \frac{1}{2}(u_{xy} + u_{yx}) = \frac{1}{2}S_6 \tag{40}$$

The coefficients of the strain tensor are subject to compatibility conditions that can be found in standard texts on elasticity theory (e.g., Landau and Lifschitz, 1959).

3. Thermodynamic Considerations

The deformation of a solid by body and surface forces results in an increase (or decrease) in the internal energy. This change can be calculated from the work done by these forces. The surface forces due to an external stress induce a deformation δu. The work δW_s will then be a product of these local deformations and forces integrated over the volume of the solid, i.e.,

$$\int_v \delta W_s \, dv = \int_v F_B \cdot \delta u \, dv = -\int_v (\nabla : \sigma) \cdot \delta u \, dv \tag{41}$$

If there is a distribution of charge within the solid defined by the net charge density ρ, then it is possible to calculate a potential ϕ from Poisson's equation

$$\nabla \cdot D = \rho \tag{42}$$

where D is the electric displacement vector. If, as a consequence of a deformation, the charge density also varies by $\delta\rho$, then the energy change will be

$$\int_v \delta W_e \, dv = \int_v \delta\rho \, \phi \, dv = \int (\nabla \cdot \delta D)\phi \, dv \tag{43}$$

As a result, the net work of deformation δW will be (in this case $\delta W = \delta W_s + \delta W_e$)

$$\int_v \delta W \, dv = -\int_v [(\nabla \cdot \sigma) \cdot \delta u - (\nabla \cdot \delta D)]\phi \, dv \tag{44}$$

Integrating the right-hand side of Eq. (44) by parts yields

$$\int_v \delta W \, dv = -\int_v \nabla \cdot (\sigma \cdot \delta u - \delta D \phi) dv + \int (\sigma : \nabla \delta u - \delta D \cdot \nabla \phi) dv \quad (45)$$

From the divergence theorem, the first integral on the right-hand side of Eq. (45) is

$$\int_v \nabla(\sigma \cdot \delta u - \phi \, \delta D) \, dv = \int_s n \cdot (\sigma \cdot \delta u - \phi \, \delta D) \, da \xrightarrow[S \to \infty]{} 0 \quad (46)$$

The disappearance of this term is due to the vanishing of the components of the right-hand side of Eq. (46), that is, the stress displacement, potential, and electric displacement. Using the definition of the electric field $E = -\nabla\phi$, and the strain $\delta S = \nabla \delta u$, then

$$\int_v \delta W \, dv = -\int_v (\sigma : \delta S + E \cdot D) \, dv \quad (47)$$

which leads to the relation

$$\delta W = -(\sigma : \delta S + E \cdot \delta D) \quad (48)$$

The differential form of the second law of thermodynamics defines the internal energy as (Landau and Lifschitz, 1959)

$$\delta U = T \, \delta S - \delta W = T \, \delta S + \sigma : \delta S + E \cdot \delta D \quad (49)$$

where T is the absolute temperature and S is the entropy. The Gibbs free energy is

$$G = U - TS - \sigma \cdot S - E \cdot D \quad (50)$$

Using the differential form of this relation and combining with Eq. (49) yields

$$\delta G = -S \, \delta T - S : \delta\sigma - D \cdot \delta E \quad (51)$$

The free energy can be expanded in a linear form in δT, δE, and $\delta\sigma$:

$$\delta G = \frac{\partial G}{\partial T}\bigg|_{E,\sigma} \delta T + \frac{\partial G}{\partial E}\bigg|_{T,\sigma} \cdot \delta E + \frac{\partial G}{\partial \sigma}\bigg|_{T,E} : \delta\sigma \quad (52)$$

which, when combined with Eq. (51), yields

$$-S = \frac{\partial G}{\partial T}\bigg|_{E,\sigma} \quad (53)$$

$$-D = \frac{\partial G}{\partial \mathbf{E}}\bigg|_{T,\sigma} \tag{54}$$

$$-\mathbf{S} = \frac{\partial G}{\partial \boldsymbol{\sigma}}\bigg|_{T,E} \tag{55}$$

From these equations, it is straightforward to show that

$$\frac{\partial S}{\partial E} = \frac{\partial \mathbf{D}}{\partial T} = \mathbf{p} \tag{56}$$

$$\frac{\partial S}{\partial \boldsymbol{\sigma}} = \frac{\partial \mathbf{S}}{\partial T} = \boldsymbol{\alpha} \tag{57}$$

$$\frac{\partial \mathbf{D}}{\partial \boldsymbol{\sigma}} = \frac{\partial \mathbf{S}}{\partial E} = \mathbf{d} \tag{58}$$

where $\mathbf{p}$ is the pyroelectric vector (a tensor of first rank), $\boldsymbol{\alpha}$ is the thermal expansion tensor (a tensor of second rank), and $\mathbf{d}$ is the piezoelectric tensor (a tensor of third rank). These tensors have components that depend not only on the symmetry of the crystal (Mason, 1950) but also on the microscopic distribution of atoms or molecules in the solid. A detailed discussion can be found in various monographs on the subject (Cady, 1969; Herbert, 1982; Lang and Glass, 1977; Mason, 1966; Nye, 1957; Tiffany, 1975).

4. Elastic Behavior and Linear Response

As pointed out earlier, the relation between the forces and the response of sensor structures is assumed to be linear. In this respect, the discussion follows the general treatment of irreversible thermodynamics (DeGroot and Mazur, 1969). Therefore the incremental strain is linear in δT, $\delta \mathbf{E}$, and $\delta \boldsymbol{\sigma}$, so that

$$\delta \mathbf{S} = \frac{\partial \mathbf{S}}{\partial T}\bigg|_{E,\sigma} \delta T + \frac{\partial \mathbf{S}}{\partial \mathbf{E}}\bigg|_{T,\sigma} \cdot \delta \mathbf{E} + \frac{\partial \mathbf{S}}{\partial \boldsymbol{\sigma}}\bigg|_{T,E} : \delta \boldsymbol{\sigma} \tag{59}$$

The only term in Eq. (59) that is undefined is $\partial \mathbf{S}/\partial \boldsymbol{\sigma}|_{T,E} = \mathbf{s}$, a fourth-rank tensor called the elastic compliance tensor. It is straightforward to define the reciprocal quantity $\partial \boldsymbol{\sigma}/\partial \mathbf{S} = \mathbf{c}$, another fourth-rank tensor called the elastic tensor. This is a form of Hooke's law linearly relating stress and strain. By using these definitions and Eqs. (57) and (58) in Eq. (59), the strain can be written as

$$\delta S = s : \delta\sigma + d \cdot \delta E + \alpha \, \delta T \tag{60}$$

In a similar fashion, the electric displacement can be expanded as

$$\delta D = \frac{\partial D}{\partial T}\bigg|_{E,\sigma} \delta T + \frac{\partial D}{\partial E}\bigg|_{T,\sigma} \delta E + \frac{\partial D}{\partial \sigma}\bigg|_{T,E} \delta\sigma \tag{61}$$

Using Eqs. (56) and (58) in Eq. (61)

$$\delta D = p \, \delta T + \epsilon \cdot \delta E + d \cdot \delta\sigma$$

where $\epsilon = \partial D / \partial E|_{T,\sigma}$ is the dielectric tensor. Finally, the linear expansion of the entropy relation yields

$$T \, \delta S = T \frac{\partial S}{\partial T}\bigg|_{E,\sigma} \delta T + T \frac{\partial S}{\partial E}\bigg|_{T,\sigma} \delta E + T \frac{\partial S}{\partial \sigma}\bigg|_{T,E} : \delta\sigma \tag{62}$$

It can be shown with Eqs. (56) and (57) that Eq. (62) is

$$T \, \delta S = \rho_M C_p \, \delta T + T(p \cdot \delta E + \alpha : \delta\sigma)$$

where ρ_M is the mass density and C_p is the specific heat at constant pressure, derived from $\rho_M C_p = \partial S / \partial T|_{E,\sigma}$.

These equations are the basis for both the piezoelectric and pyroelectric sensors. If T, E, and/or σ are time-varying functions, then the strain, charge state, and heat energy of the sensor materials will also be time-varying.

5. Dielectric Characteristic Related to Piezoelectric and Pyroelectric Structures

One of the basic relations in physics is the continuity equation

$$\nabla \cdot J + \frac{\partial \rho}{\partial t} = 0 \tag{63}$$

where J is the current density. Using Poisson's equation, Eq. (42), Eq. (63) can be rewritten in the form

$$\nabla \cdot \left(J + \frac{\partial D}{\partial t} \right) = 0 \tag{64}$$

Integrating this expression yields

$$J = -\frac{\partial D}{\partial t} + \text{const} \tag{65}$$

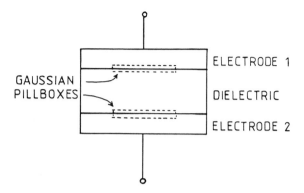

Fig. 4. Capacitive structure illustrating the Gaussian pillboxes needed to define the electrostatic boundary conditions in Laplace's equation in the dielectric.

If an electric field is applied to a solid, not only the electric displacement vector **D** but also the current will change. At steady state $\partial \mathbf{D}/\partial t = 0$ and the constant in Eq. (65) will be $\sigma_e \mathbf{E}$ (i.e., Eq. (65) reduces to Ohm's law), so that

$$\mathbf{J} = -\frac{\partial \mathbf{D}}{\partial t} + \sigma_e \mathbf{E} \tag{66}$$

where σ_e is the electrical conductivity of the material. In the case of most piezo- and pyroelectric materials, σ_e is so small that, over the times of interest, it is safe to set this last term equal to zero.

If a capacitive structure is employed as shown in Fig. 4, then it is necessary to examine the charge state of the metal electrode plates. If a Gaussian pillbox is inscribed through the surface of the dielectric and metal, then from Poisson's equation

$$\int_{v,1} \mathbf{\nabla} \cdot \mathbf{D} \, dv = \int_{v,1} \rho \, dv = \int_{S,1} Q(1) \, da = \int_{S,1} \mathbf{D}(1) \cdot \mathbf{n} \, da \tag{67}$$

$$\int_{v,2} \mathbf{\nabla} \cdot \mathbf{D} \, dv = \int_{v,2} \rho \, dv = \int_{S,2} Q(2) \, da = - \int_{S,2} \mathbf{D}(2) \cdot \mathbf{n} \, da \tag{68}$$

The minus sign in Eq. (68) takes into account that the unit vector **n** always points away from the surface. Since **D** is in the same direction at 1 and 2, the negative sign is essential. Therefore, in the metal

$$Q(1) = \mathbf{D}(1) \cdot \mathbf{n} \tag{69}$$

$$Q(2) = -\mathbf{D}(2) \cdot \mathbf{n} \tag{70}$$

In the dielectric, ρ is zero and

$$\frac{d(\mathbf{D} \cdot n)}{dx} = 0 \tag{71}$$

or, inside the material,

$$Q = \mathbf{n} \cdot \mathbf{D}(x) \tag{72}$$

where Q is an arbitrary constant derivable from the boundary conditions listed in Eq. (67) and (68). Thus

$$Q = \mathbf{n} \cdot \mathbf{D}(1) = Q(1) \tag{73}$$

$$Q = \mathbf{n} \cdot \mathbf{D}(2) = -Q(2) \tag{74}$$

Since

$$Q(1) = -Q(2) \tag{75}$$

this establishes charge conservation in a capacitive structure.

6. The Wave Equation

The excitation of waves in piezoelectric structures is rather complex because of the tensorial relations between the piezoelectric tensor and the elasticity tensor. Applying an electric field to the capacitor induces a strain. The relaxation of the resulting stress induces a charge which, when applied to an oscillator circuit, induces a resonant response. There will be no attempt here to discuss all the vibrational modes in depth. Rather, an illustrative example based on the simple longitudinal vibrational mode will be presented. As shown schematically in Fig. 5, the surfaces in the y and z axes are stress-free. Therefore, there are no shear components in the stress and the tensile stresses are also zero. Because the surfaces of the crystal are connected by a conducting electrode, the field components in the y and z directions are also zero. As a result, the stress tensor can be written in matrix notation as

$$\delta\boldsymbol{\sigma} = \begin{vmatrix} \sigma_{xx} & 0 & 0 \\ 0 & 0 & 0 \\ 0 & 0 & 0 \end{vmatrix} \tag{76}$$

and the electric field vector is

$$\delta\mathbf{E} = \begin{pmatrix} 0 \\ 0 \\ E_z \end{pmatrix} \tag{77}$$

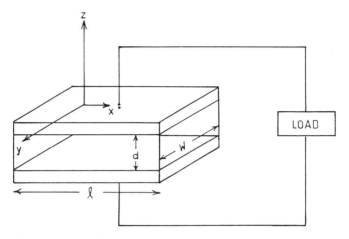

Fig. 5. Schematic layout of the piezoelectric structure. As a result of a strain in the capacitor, a displacement current flows through the load. This results in a signal that can be used to vary an electric field on the capacitor. This in turn induces a strain, closing the feedback loop for a piezoelectric oscillator.

Note that the surfaces at $x = 0$ and $x = l$ in Fig. 5 are free surfaces. As a result, σ_{xx} will be zero at $x = 0$ and at $x = l$. The piezoelectric tensor for quartz in this case is

$$\mathbf{d} = \begin{bmatrix} 0 & 0 & d_{11}\cos\gamma_1\cos^2\alpha_1 + d_{14}\cos\gamma_1\cos\alpha_2\cos\alpha_3 \\ 0 & 0 & 0 \\ 0 & 0 & 0 \\ 0 & 0 & 0 \\ 0 & 0 & 0 \\ 0 & 0 & 0 \\ 0 & 0 & 0 \\ 0 & 0 & 0 \\ 0 & 0 & 0 \end{bmatrix} \tag{78}$$

where α_i and γ_j are direction cosines of the displaced coordinate system about the original axis of the crystal (Fig. 6).

The only force is along the x axis, so the strain component in the x direction is [from Eq. (60)]

$$S_{xx} = s_{11}\sigma_{xx} + d_{13}E_z = S_{11} \tag{79}$$

where d_{13} is the nonzero term in the matrix form of Eq. (78). Rearranging terms and using Eq. (35) for S_{11},

$$\sigma_{xx} = \frac{S_{xx} - d_{13}E_z}{s_{11}} = \frac{\partial u/\partial x - d_{13}E_z}{s_{11}} \tag{80}$$

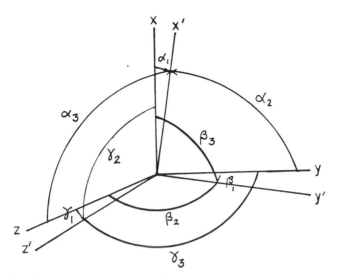

Fig. 6. Coordinate axes showing the angles connecting the rotated (x',y',z') system to the original (x,y,z) system.

The element of force on the element e^{ζ} volume, according to Newton's second law and Eq. (24), is

$$d\mathbf{F}_x = -\nabla \cdot \boldsymbol{\sigma} \, dv = \rho_M \frac{d^2u}{dt^2} \, dv \tag{81}$$

Using Eq. (80) for σ_{xx},

$$\nabla \cdot \boldsymbol{\sigma} = \frac{\partial \sigma_{xx}}{\partial x} = \frac{-1}{s_{11}} \left(\frac{\partial^2 u}{\partial x^2} - d_{13} \frac{\partial E_z}{\partial x} \right) \tag{82}$$

As pointed out before, E_z is not a function of the lateral x position because the capacitor electrodes are equipotentials (see Fig. 5). Therefore, $\partial E_z / \partial x = 0$ and

$$\rho_M \frac{d^2u}{dt^2} = \frac{1}{s_{11}} \frac{\partial^2 u}{\partial x^2} \tag{83}$$

This is a wave equation, and it has a simple solution of the form

$$u = u(x) \, e^{j\omega t} \tag{84}$$

where ω is the angular frequency of the electric field excitation frequency $\delta \mathbf{E} \, e^{j\omega t}$. Note that the velocity of sound in the dielectric is

$$V = \frac{1}{\sqrt{\rho_M s_{11}}} \tag{85}$$

It is simple to show that the amplitude of the longitudinal modes is given by

$$u = d_{13}E_z \left\{ \frac{\sin[m(l - 2x)/2]}{\cos(ml/2)} \right\} \tag{86}$$

where $m = \omega/V$ and l is the length of the piezoelectric crystal. Use is made of the boundary condition $\sigma_{xx} = 0$ at $x = 0$ and $x = l$ to obtain Eq. (86).

From the current equation, Eq. (66), and the linear form for $\delta \mathbf{D}$, Eq. (61), the current in the circuit shown in Fig. 5 will be

$$\mathbf{J} = -\frac{\partial \mathbf{D}}{\partial t} = -\left(\boldsymbol{\epsilon} \cdot \frac{\partial \mathbf{E}}{\partial t} + \mathbf{d} : \frac{\partial \boldsymbol{\sigma}}{\partial t} \right) = -j\omega(\boldsymbol{\epsilon} \cdot \mathbf{E} + \mathbf{d} : \boldsymbol{\sigma}) \tag{87}$$

From the boundary conditions, only the $\bar{z}$ components of the electric field exist and $\boldsymbol{\sigma}$, according to Hooke's law, must be in the form

$$S_{11} = s_{11}\sigma_{11} \tag{88}$$

From Eq. (78), $\mathbf{d}$ reduces to d_{13} and the current density becomes

$$J_z = -j\omega \left(\varepsilon_z \varepsilon_0 E_z + \frac{d_{13}}{s_{11}} \frac{\partial u}{\partial x} \right) \tag{89}$$

This result makes use of Eq. (35) for S_{11}. The wave equation solution for $u(x)$, Eq. (86), can now be introduced into Eq. (89) and the current density obtained by integrating over the electrode surface:

$$I = \int_0^l J_z(x) W \, dx = -j\omega A \left[\varepsilon_z \varepsilon_0 + \frac{d_{13}^2 \tan(ml/2)}{ml/2} \right] E_z \tag{90}$$

where ε_z is the diagonal component of the dielectric tensor in the z direction, ε_0 is the permittivity of free space, W is the width of the piezoelectric, and A is the area Wl. Since $V = -E_z d$, $i = JA$, and d is the thickness of the piezoelectric crystal, the impedance $\bar{\mathfrak{z}}$ of the crystal becomes

$$\frac{1}{\bar{\mathfrak{z}}} = \frac{j\omega A}{d} \left\{ \varepsilon_z \varepsilon_0 + \frac{d_{31}'^2}{s_{11}} \left(\frac{\tan(ml/2)}{ml/2} \right) \right\} \tag{91}$$

When $\tan(ml/2) = \infty$ the impedance vanishes, corresponding to a resonance in the circuit (Fig. 5). The resonance frequency arises from the condition that

$$\frac{ml}{2} = \frac{\pi}{2} \tag{92}$$

which corresponds to

$$f_R = \frac{V}{2l} = \frac{1}{2l}(\rho_M s_{11}) \tag{93}$$

where f_R is the resonance frequency of the system. The resonance frequency can also be expressed in terms of the total mass M of the crystal, including the mass of the electrodes and any chemically sensitive layer on the electrodes. Replacing ρ_M by M/V yields

$$f_R \equiv \frac{1}{2\sqrt{(l/dW)MS_{11}}} \tag{94}$$

Any change in M will induce a change in f_R, that is,

$$\frac{\delta f_R}{f_R} = -\frac{\delta M}{2M} \tag{95}$$

C. PIEZOELECTRIC GRAVIMETRIC SENSOR

The piezoelectric gravimeter, in its simplest form, is a capacitor with a dielectric material having a preferred axis normal to the capacitor plates; the device can be excited by an electric field at its mechanical resonance frequency, given by Eq. (93). Since the resonance frequency depends on the total mass of the gravimeter, any small change in mass will shift the resonance frequency as indicated in Eq. (95). Suppose the gravimeter is constructed as shown in Fig. 7. Then the total mass will be

$$M = M_{PG} + m + \delta m \tag{96}$$

where M_{PG} is the mass of the piezoelectric gravimeter in the absence of the mass δm due to the products of reaction with the chemically sensitive layer and m, the mass of the chemically sensitive layer. From Eq. (95),

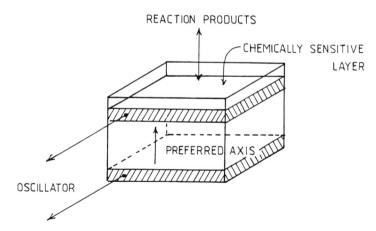

Fig. 7. Setup for the piezoelectric gas detector. The chemically sensitive layer reacts with the species to be detected, producing reaction products.

the shift in the resonance frequency will be

$$\frac{\Delta f_R}{f_R} = - \frac{\delta m}{2M} = - \frac{\delta m}{2M_{PG}}\left(1 - \frac{m + \delta m}{M_{PG}}\right) \qquad (97)$$

Since $(M + \delta M)/M_{PG} \ll 1$, any change in the mass of the chemically sensitive layer will be directly observable as a change in Δf_R. From Eq. (93), it is easy to show that the resonance frequency is

$$f_R = \frac{1}{2}\left(\frac{l}{dW}M_{PG}s_{11}\right)^{-1/2} \qquad (98)$$

that is, to a first approximation f_R is determined by the mass of the piezo-electric crystal if the slight correction for the mass of the electrodes and the chemically sensitive layer can be neglected. This is not only the basis for piezoelectric gravimeters for chemical detection, it also explains the principle for the piezoelectric film thickness monitors (Behrndt and Love, 1962).

The quality factor, defined as the ratio of the bandwidth of the crystal oscillator to the resonance frequency, can be quite high for a quartz crystal. As an illustration, a thermostatically controlled quartz crystal ($\Delta T \simeq \pm 0.1°C$) can be used as the control element in an oscillator circuit to give a frequency that has a bandwidth of 1 Hz at a center frequency of 10 MHz. Assuming that the minimum measureable frequency shift is of the order of 1 Hz, then δm can be calculated from Eq. (95) (assuming that M_{PG} is in the 10-mg range) to be

$$\delta m \simeq 2 \times 10^{-9} \text{ g} = NM_{mol} \qquad (99)$$

Assuming a median molecular weight of 50 g per mole or 10^{-22} g per molecule, this value of δm corresponds to a surface molecular density N of 2×10^{14} cm^{-2}. What is interesting to note is that this is roughly of the order of 0.1 monolayer. While not the ultimate in chemical sensitivity, it is certainly adequate for many purposes.

Although almost any piezoelectric material could be employed, α-quartz is most commonly used because of its highly desirable physico-chemical properties and its ready availability. This form of quartz is hex-agonal and the piezoelectric coefficients along different crystallographic directions can have widely varying temperature coefficients (Mason, 1950). Two cuts widely used for making oscillator crystals are shown in Fig. 8. The most commonly used cut for piezogravimetry is the AT cut. It has the lowest piezoelectric constant-temperature coefficient and is most readily available commercially as a 9-MHz crystal. The dimensions are in the range 1–1.5 cm with thicknesses of 0.1–0.15 mm. AT-cut crystals will

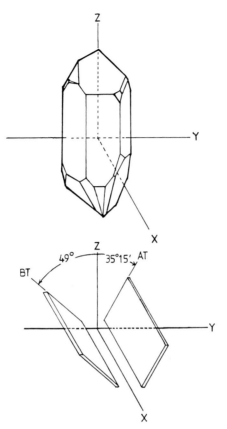

Fig. 8. Schematic drawing of a quartz crystal showing the standard axes. The relative angles for the AT and BT cuts are shown below. These cuts possess low temperature coefficients of oscillation frequency.

vibrate in the thickness shear mode as indicated in Fig. 9. While this is mathematically a different mode than the longitudinal mode discussed above, the principles are the same. An exploded view of a piezoelectric crystal thickness monitor suitable for piezoelectric gravimetry is shown in Fig. 10; a schematic of the circuitry used to monitor the rate of change of the adsorbed (or deposited) layers is presented in Fig. 11.

1. Sorption Detection

   Use of a piezogravimetric sensor for gas detection was first reported by King *et al.* (1968). The initial materials detected are listed in Table I. In a much more comprehensive review, Hlavay and Guilbault (1980) listed a

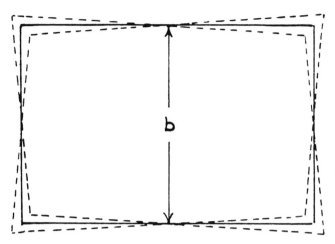

Fig. 9. Type of shear oscillation arising in the AT-cut crystal. The result of this shear oscillation (shown as dashed lines) is a spatial variation of the piezoelectric mass sensitivity over the face of the crystal.

TABLE I

TYPES OF DETECTOR COATINGS ON PIEZOELECTRIC MASS
DETECTORS FOR VARIOUS VAPORS AND GASES[a]

Detector characteristic	Coating material
Hydrocarbon detection, nonselective for compound type	Squalane
	Silicone oil
	Apiezon grease
Selective detection of polar molecules such as aromatic, oxygenated, and unsaturated compounds	Polyethylene glycol
	Sulfolane
	Dinonyl phthalate
	Aldol-40
	Tide (alkyl sulfonate)
Water vapor	Silica gel
	Molecular sieve
	Alumina
	Hygroscopic polymers[b]
Hydrogen	Lead acetate
	Metallic silver
	Metallic copper
	Anthraquinone disulfonic acid

[a] After King et al., 1968.
[b] Natural resins, glues, cellulose derivatives, and synthetic polymers.

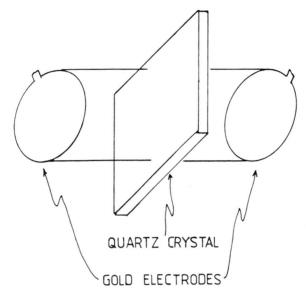

QUARTZ CRYSTAL

GOLD ELECTRODES

Fig. 10. Exploded view of the electrodes on a quartz oscillator crystal.

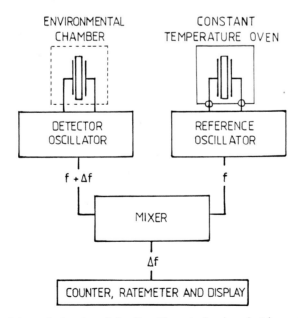

ENVIRONMENTAL CHAMBER

CONSTANT TEMPERATURE OVEN

DETECTOR OSCILLATOR

REFERENCE OSCILLATOR

$f + \Delta f$

f

MIXER

Δf

COUNTER, RATEMETER AND DISPLAY

Fig. 11. Schematic drawing of circuit and layout of a piezoelectric gravimeter.

wide range of coatings that have been used to detect numerous gas species. A partial listing is given in Table II.

The general advantage of the piezogravimetric sorption detector is that a single electronic structure (the piezogravimetric crystal) provides a well-defined substrate for a wide variety of coatings. This was pointed out in the Introduction and the data in Tables I and II make it clear how advantageous this approach can be; however, there is a problem in using the piezogravimetric vehicle. In general, this device operates at the lowest mechanical frequency of the crystal and involves the coherent motion of the entire crystal. Under these circumstances it can function only in a "one species, one crystal" manner. This is particularly undesirable because the selectivity of the various coatings can hardly be described as ideal. Recently, Hirschfeld proposed the use of a higher order transverse mode excitation to allow monitoring of multiple species on a single oscillator chip at various modes (T. Hirschfeld, private communication, 1983). Unpublished reports by Hirschfeld suggest that this may be a very useful approach.

2. Piezothermogravimetry (PTG)

Recently, Henderson and co-workers (1982) extended the basic concept of piezogravimetry to include the effect of a thermal ramp. Ramping the temperature varies the evaporation rate from a sample deposited on the PTG structure. As mentioned in Section II,C, the AT-cut crystal is selected for its relatively high temperature stability. However, if the crys-

TABLE II

TYPICAL COATINGS USED IN GAS DETECTORS[a]

Detected molecule	Typical coating	Interferences
SO_2	Quadrol	NO_2, H_2O
	Triethanolamine	NO_2, H_2O
H_2O	LiCl	
	SiO_x	
NH_3	Ucon 75-H-90,000; LB 300x	H_2O, organics
	Ascorbic acid/ascorbic acid	
	$AgNO_3$	
Diisopropylmethyl phosphonate	$FeCl_3$	
(DIMP) and related	Paraoxon	H_2O
compounds		
H_2S	Soot from chlorobenzoic acid	
HCl	Triphenylamine	

[a] After Hlavaty and Guilbault, 1980.

tal was thermally ramped, a significant change in the frequency of the oscillator would be expected. Studies on a 5.0-MHz, AT-cut, quartz crystal from 0 to 400°C showed a variation of 1020 Hz. The only way to handle this kind of frequency variation is to employ a computer to control both the thermal ramp and the data acquisition and processing.

The use of computer data and analysis was vital. The experimental apparatus included two quartz oscillator crystals mounted side by side (one a temperature monitor, the other the PTG) on a metal (copper) strip heated by a soldering gun in one configuration and a combination of a tube furnace and thermocouple temperature monitor in the other. These configurations are shown schematically in Figs. 12 and 13. There did not appear to be any feedback control of the temperature of the PTG or computer-controlled input to the heaters.

The measurement procedure itself was quite standard. The PTG crystal was incorporated into a crystal-controlled oscillator (CCO); a counter read the frequency of the CCO and transferred the data to a microcomputer. The temperature of the PTG crystal was monitored separately either with the other quartz crystal mounted in an oscillator or with a thermocouple. In the measurements reported here, most of the data were from the thermocouple measuring system. After careful mass, temperature, and gas flow calibration, it was possible to obtain a precise value for the

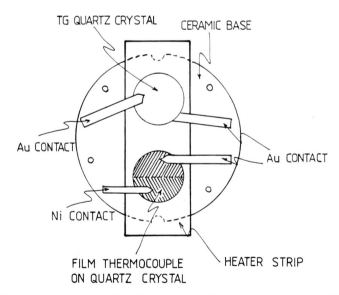

Fig. 12. Schematic layout of the piezothermogravimetric detector system with a strip heater. Two standard AT-cut crystals are used, one as the detector and the other to monitor the temperature with a thin-film thermocouple. (Adapted with permission from Henderson *et al.*, 1982. Copyright 1982 American Chemical Society.)

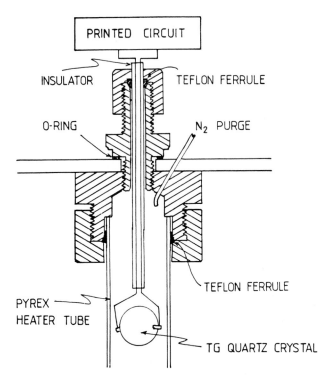

Fig. 13. Schematic layout of the furnace-heated, piezothermogravimetric detector. The printed circuit contains the oscillator circuit. (Adapted with permission from Henderson *et al.*, 1982. Copyright 1982 American Chemical Society.)

mass loss as a function of temperature. The software analysis of the data was carried out directly with the microcomputer. A least-squares fit to the temperature dependence of the frequency of the PTG oscillator was subtracted from the input data and other corrections for gas flow and initial mass were made. An example of the resultant data is shown in Fig. 14.

The significant aspect of this study was the use of a small, computer-monitored chip for a chemical evaluation. As will be seen in Section II,D, this procedure is even more vital for pyroelectric gas sensor measurements than for the PTG study. Nevertheless, the data demonstrate that submicrogram quantities of material can be analyzed with these structures.

D. PYROELECTRIC GAS ANALYSIS

Pyroelectricity arises from the temperature dependence of the spontaneous polarization of certain noncubic crystals and polymeric materials.

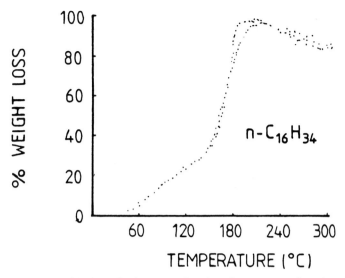

Fig. 14. Percent weight loss of n-$C_{16}H_{34}$ as a function of temperature. Sample weight was 1.4 μg, and nitrogen gas flowed over the sample at 130 ml/min. Rate of temperature increase was 100 °C/min; data sampling rate was 10 points/s. There is no explanation for the increase in weight for $T \geq 200$°C. (Adapted with permission from Henderson *et al.*, 1982. Copyright 1982 American Chemical Society.)

Unlike piezoelectric crystals, the pyroelectric material must have a spontaneous polarization even at zero electric field. Phenomenologically, the pyroelectric material responds to temperature change the way a piezoelectric crystal responds to stress. Assuming that the pyroelectric material is a respectable insulator, the change in polarization will induce current flows external to the crystal.

The change in the displacement vector for a material with a single pyroelectric axis can be obtained from Eq. (61) by spatial averaging of $\delta \mathbf{D}$,

$$\langle \delta \mathbf{D} \rangle = \langle \mathbf{D} \rangle - \mathbf{D}_0 = \mathbf{p}(T)\langle \delta T \rangle + \varepsilon_p \varepsilon_0 \langle \delta \mathbf{E} \rangle \tag{100}$$

where ε_p is the dielectric tensor component along the pyroelectric axis and the brackets refer to a volume average

$$\langle \delta y \rangle = \frac{1}{v} \int_v \delta y \, dv = \langle y \rangle - y_0 \tag{101}$$

Using this relation,

$$\langle \delta T \rangle = \langle T \rangle - T_0 \tag{102}$$

$$\langle \delta E \rangle = -\frac{1}{d}(V - V_0) \tag{103}$$

where $V - V_0$ is the potential drop across the pyroelectric capacitor. The vectors will be treated as scalars henceforth. It then follows that

$$\frac{\varepsilon_p \varepsilon_0}{d} = C_p \tag{104}$$

so that the charge density will be

$$\langle \delta D \rangle = Q = -C_p(V - V_0) + p(T)(\langle T \rangle - T_0) \tag{105}$$

It is implicitly assumed that p is very weakly dependent on temperature over the temperature range of interest, that is, from an initial temperature T_0 to a maximum temperature $T_{\max}$. The charge density on the pyroelectric condenser will vary with time as the temperature varies with time. The total pyroelectric current will be the integral of the current density over the capacitor area, that is,

$$i_p = \int_A \frac{dQ}{dt}\, da = -C_p A \frac{dV}{dt} + p_0 A \frac{d\langle T \rangle}{dt} \tag{106}$$

This current can be viewed as arising from a constant-current generator in the equivalent circuit shown in Fig. 15. The voltage at the amplifier in Fig. 15 will then be

$$V = i_p R = -RCA \frac{dV}{dt} + p_0 RA \frac{d\langle T \rangle}{dt} \tag{107}$$

where

$$R = \left(\frac{1}{R_p} + \frac{1}{R_A}\right)^{-1} \tag{108}$$

$$C = (C_p + C_A) \tag{109}$$

$$RC = \tau \tag{110}$$

R_p is the resistance of the pyroelectric capacitor (typically 10^{10}–10^{14} ohms), R_A is the input resistance of the first stage of the detector amplifier, and C_A is the input capacitance of the amplifier. The measured voltage will then be

$$V = \frac{pA \exp(-t/\tau)}{C} \int_0^t \exp(t'/\tau) \left\langle \frac{dT}{dt'} \right\rangle dt' \tag{111}$$

If the temperature varies sinusoidally at a frequency ω, the voltage will be

$$V(\omega) = \frac{j\omega p RA \langle T \rangle}{1 + j\omega\tau} \tag{112}$$

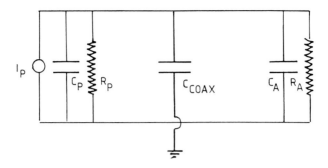

Fig. 15. Equivalent circuit for a pyroelectric crystal and associated amplifier. The subscript p refers to the pyroelectric device and A refers to the preamplifier; C_{COAX} is the shunt capacity of the coaxial connection between the pyroelectric device and the preamplifier.

The temperature distribution can be determined by solving the well-known Fourier heat equation subject to appropriate boundary conditions (Carslaw and Jaeger, 1959).

In the case of one-dimensional heat flow, the Fourier heat equation becomes

$$\frac{\partial T}{\partial t} = D_T \frac{\partial^2 T}{\partial z^2} \tag{113}$$

where D_T is the thermal diffusivity, defined as the ratio of the thermal conductivity κ (W/cm-°C) to the volume heat capacity C_v (J/cm³-°C). Using Eq. (113) in Eq. (101) yields

$$\left\langle \frac{\partial T}{\partial t} \right\rangle = \frac{\kappa}{C_v} \left\langle \frac{\partial^2 T}{\partial z^2} \right\rangle = \frac{1}{C_v d} [H(0, t) - H(d, t)] \tag{114}$$

where $H(i, t)$ is the heat flow across the interfaces defined in Fig. 16. Since $H(0, t) - H(d, t) = \Delta H(t)$ is the net heat flowing into the pyroelectric, substituting Eq. (114) into Eq. (112) leads to

$$V = \frac{A p_0 \exp(-t/\tau)}{C C_v} \int_0^t \Delta H(t') \exp\left(\frac{t'}{\tau}\right) dt' \tag{115}$$

If τ is small compared to the characteristic thermal flow time constants, then it is relatively easy to show that

$$V = \frac{p_0 R A}{C_v d} \Delta H(t) \tag{116}$$

The voltage will depend on the value of the RC time constant of the circuit in time intervals corresponding to $0 \leq t \leq 3\tau$. Since typical values of τ are

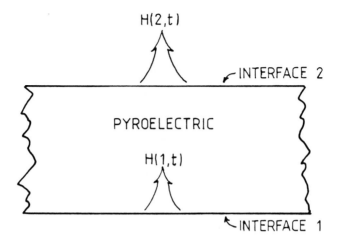

Fig. 16. Heat flow schematic in the pyroelectric gas sensor. Interface 1 refers to the substrate heater and interface z to the gas-sensitive layer.

of the order of 10 ms, Eq. (116) should hold as long as the thermal time constant associated with $\Delta H(t)$ is $\geqslant 10$ ms. Another way to visualize the thermal response of the system is to assume that the thermal time constant of the system is τ_T. The thermal response can be shown to vary as

$$\langle T \rangle = \frac{\Delta T}{1 + j\omega\tau_T} \tag{117}$$

where ΔT is the peak temperature excursion of the sinusoidal temperature variation. Introducing Eq. (117) into Eq. (112) yields

$$V(\omega) = \frac{pRA\,\Delta T\,j\omega}{(1 + j\omega\tau_T)(1 + j\omega\tau)} \tag{118}$$

At low enough frequencies, $V(\omega) \rightarrow 0$. There will be a range where either $|j\omega\tau| \gg 1$ or $|j\omega\tau_T| \gg 1$ and where the other term is small, that is,

$$V(\omega) \simeq \begin{cases} \dfrac{pRA\,\Delta T}{\tau_T} & \text{if} \quad \tau_T > \tau \\[4mm] \dfrac{pRA\,\Delta T}{\tau} & \text{if} \quad \tau > \tau_T \end{cases} \tag{119}$$

Finally, at high enough frequencies, $V(\omega) \rightarrow 0$. The highest value of $V(\omega)$ will occur when $\tau_T = \tau$.

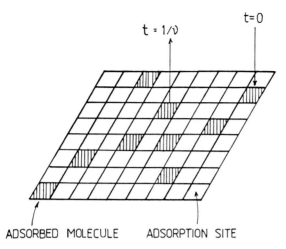

Fig. 17. Schematic of the surface of the gas-sensitive layer showing the adsorbed molecules (hatching) on the adsorption sites (blank). A molecule adsorbs at time $t = 0$ and desorbs at time $t = 1/v$.

1. Response Characteristics of the Pyroelectric Gas Analyzer (PGA)

Two factors will control the response of the PGA. The first is the heat carried away by desorption of the gases on the surface. For convenience, only one gas is assumed to be present, although the case with two or more gases can be solved. The second factor is the nondesorptive heat losses, such as radiative losses or thermal conduction through the contacting wire leads. The latter is assumed to be negligible.

For this discussion, a simple first-order hyperbolic thermally programmed desorption process is assumed. The surface coverage θ is the ratio of the surface density of adsorbed molecules, N, to the surface density of adsorption sites, N_0. The various parameters are illustrated in Fig. 17. The rate of desorption is then

$$vd = v \exp\left[\frac{Q_a}{kT(t)}\right] \qquad (120)$$

where v is the natural frequency of the adsorbed molecule, Q_a is the binding energy (the quantity that provides the fundamental discrimination between the adsorbed species), k is the Boltzmann constant, and $T(t)$ is the temperature, which is subject to a time variation. One of the simpler variations is the hyperbolic heating rate

$$\frac{1}{T(t)} = \frac{1}{T_0} - at \qquad (121)$$

which leads to a rate of desorption

$$\nu_d = \nu_0 \exp\left[\left(\frac{Q_a a}{K}\right) t\right] \tag{122}$$

where $\nu_0 = \nu_d(T_0)$.

For convenience, the first-order desorption rate equation is used:

$$-\frac{dN}{dt} = \nu_d N \tag{123}$$

which can be rewritten in the dimensionless form

$$\frac{d\theta}{dy} = \theta \, e^{xy} \tag{124}$$

where

$$x = \frac{Q_a a}{K} \nu_0 \tag{125}$$

$$y = \nu_0 t \tag{126}$$

$$\theta = \frac{N}{N_0} \tag{127}$$

This has a solution

$$\theta = \exp\left(\frac{1 - e^{xy}}{x}\right) \tag{128}$$

and the desorption rate Γ_d can be expressed as

$$\frac{\Gamma_d(t)}{\nu_0 N_0} = \frac{d\theta}{dy} = \exp\left(\frac{1 - e^{xy}}{x}\right) \tag{129}$$

A plot of $\Gamma_d(t)$ as a function of time for different values of Q_a is shown in Fig. 18. The heat loss associated with dn/dt will depend on the distribution of desorption energies $g(Q_a)$. The total heat loss will be an integral over this distribution.

$$\Delta H(t) = \frac{\int \Gamma_d(t, Q_a) g(Q_a) \, dQ_a}{\int g(Q_a) \, dQ_a}$$

$$= \frac{N_0 \left(\frac{k}{a}\right)^2 \nu_0^3 \int_{x_0}^{x} \exp[xy + (1 - e^{xy})x] g(x) \, dx}{\int_{x_0}^{x} g(x) \, dx} \tag{130}$$

In addition to the desorptive energy loss, there is also a radiative heat loss term that will increase as T^4. This heat loss is not discussed here. It should

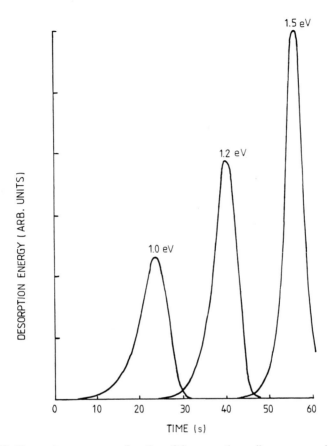

Fig. 18. Desorption energy as a function of time on a thermally programmed desorption structure. Initial temperature was 300 K, final temperature 600 K, and hyperbolic heating rate 2.77×10^{-5} K/s. The parameter is the binding energy Q_a defined in the desorption rate as $\nu_d = \nu \exp(Q_a k T(t))$. The vibrational frequency ν was chosen to be 10^{13} Hz.

be noted that a variety of chemical reactions can take place on the surface of the adsorber. If the reaction is exothermic it will produce additional heat, and if endothermic it will absorb heat. The sign of the voltage response will be an indication of whether the process is endothermic or exothermic.

So far there has been no mention of the techniques for varying the temperature of the PGA according to a prescribed $T(t)$. The straightforward procedure that has been used is to take a small tube furnace and place the temperature under the control of a temperature programmer (A. D'Amico, personal communication, 1983). Another and potentially more powerful approach is to integrate the heater directly with the pyroelectric structure (Young, 1983). Such a monolithic structure would have a response characteristic that would be equal to the PGA chip time constant. Since this is of the order of 100 m, it is possible to carefully control the temperature–time profile to obtain a predetermined $T(t)$.

2. Device Fabrication

The only pyroelectric gas sensors studied to date have been prepared from Czochralski grown, z-cut, single-crystal $LiTaO_3$ wafers, typically 230 μm thick. These 25-mm-diameter wafers were oriented with the pyroelectric axis perpendicular to the plane of the surface. The wafers were sliced from a poled boule. No further polishing of the wafers was needed since the saw marks were negligible.

The device fabrication steps are basically the same as those used in silicon device photolithography. The wafers are initially cleaned in acetone and ethanol to remove excess grease and organic compounds. They are then blown dry in a stream of N_2 from liquid nitrogen boil-off (blow-drying in N_2). They are cleaned in aqua regia and rinsed thoroughly in deionized water, followed by blow-drying in N_2. The wafers are then mounted in a vacuum evaporation system and a 300-nm gold layer is evaporated to form the electrodes on the measuring surface.

The electrodes themselves are fabricated by the standard photolithographic procedures described in Chapter 3. Positive resist is used in this photolithographic processing. The two patterns used at the University of Pennsylvania are shown in Fig. 19,b and c. In contrast, a single electrode

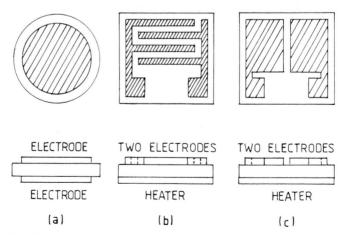

Fig. 19. Schematic of pyroelectric gas analyzer structure: (a) single-electrode detector; (b) interdigited electrode structure; (c) planar configuration.

structure has been used by D'Amico and co-workers (A. D'Amico, private communication, 1983) (Fig. 19a). After the photolithographic pattern has been developed, the excess gold is removed in dilute aqua regia. After another careful cleaning to remove excess impurities, the wafer is reversed and a 10-nm chromium film is evaporated onto the back, followed by subsequent evaporation of 400 nm of NiCr. A Cr–Au contact for the heater is then vacuum-evaporated onto the NiCr film through mechanical masks. The wafer is then mounted on a standard microscope slide with Apiezon wax and sliced with a diamond wire saw to the desired sizes. The PGA chips are released by dissolving the wax in benzene, followed by careful rinsing in acetone and alcohol. Leads are attached by gold ultrasonic ball bonding. The wire is approximately 50 μm in diameter. A schematic drawing of the PGA structure is shown in Fig. 20, and photographs of two different electrode geometries are shown in Fig. 21,a and b. The chips are mounted on a ceramic holder, as shown in Fig. 22, after which the gas-sensitive material is placed in contact with one of the two electrodes. A completed assembly with activated charcoal as the sorber is shown in Fig. 23.

3. Measurement Procedures and Equipment

As indicated in Eq. (116), the output voltage from a single plate of a PGA is directly proportional to the *net* heat flow into the pyroelectric structure. If the reference electrode contains no chemically sensitive material, then its heat loss can be viewed as a reference state for the measurement. The additional heat loss (or gain) associated with interaction processes on the chemically sensitive material would stand out clearly in

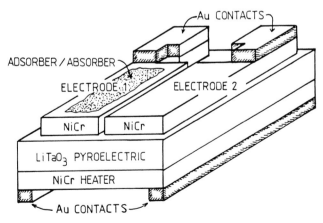

Fig. 20. Schematic of pyroelectric gas analyzer showing details of construction.

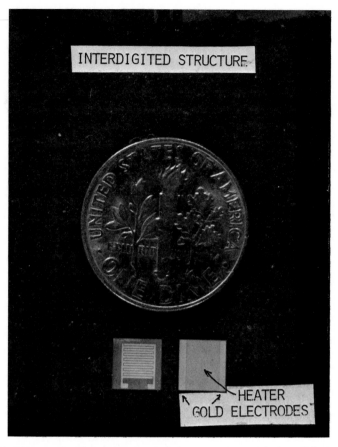

Fig. 21. Photographs of pyroelectric gas analyzer structures: left, interdigited structure; right, planar structure.

a differential measurement. In the schematic drawing of Fig. 20 there are two net heat fluxes for the two electrodes, ΔH_1 and ΔH_2, respectively. One of these net heat fluxes is shown schematically in Fig. 16. The differential voltage with respect to ground is, from Eq. (117),

$$\Delta V = \frac{ApR}{dC_v}(\Delta H_1 - \Delta H_2) = \frac{ApR}{dC_v}\delta H_{12} \qquad (131)$$

Since the heat flow into the pyroelectric material from the heater is the same for both electrodes, the difference voltage must be due to the differences in the heat loss, δH_{12}. To estimate the sensitivity of the pyroelectric detector, some simplifying assumptions are made. The first assumption limits the noise voltage to that generated by the load resistor, R_L. If the

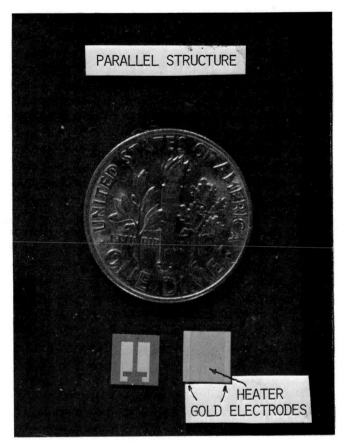

Fig. 21. (*Continued*)

signal-to-noise ratio (S/N) is unity, then Eq. (131) will approximate the Johnson noise, or

$$V_N = \sqrt{4kTR_L\,\Delta f} \cong \frac{ApR_L}{C_v d}\,\delta H_{12} \tag{132}$$

The signal is generated by heat released when the gas molecules suddenly desorb as the temperature is raised at a hyperbolic rate.

Assuming a single gas (for simplicity), the maximum in Eq. (129) will be

$$\left.\frac{d\theta}{dy}\right|_{\max} = x\,e^{(1/x-1)} \tag{133}$$

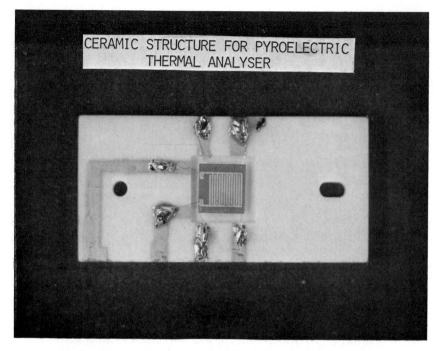

Fig. 22. Mounted but uncoated interdigited pyroelectric gas analyzer.

which leads to a value for $\Gamma_d|_{max}$ of

$$\Gamma_d\bigg|_{max} = \frac{N_0 Q_a a}{k} \exp\left(\frac{k\nu_0}{Q_a a} - 1\right) \tag{134}$$

Because only one gas is present, one adsorption site substrate type is assumed here,

$$\Delta H\bigg|_{max} = Q_a \Gamma_{d_{max}} = \frac{N_0 Q_a^2 a}{k} \exp\left(\frac{k\nu_0}{Q_a a} - 1\right) \tag{135}$$

Using the values in Table III, $\Delta H|_{max}$ is estimated to be 20 μW/cm² for a monolayer desorption. If N_0 were increased by two or three orders of magnitude, $\Delta H|_{max}$ would increase correspondingly. Using Eq. (132), the minimum heat signal for unity S/N will be

$$\Delta H_i\bigg|_{min} = \frac{C_v d}{A p_0} \sqrt{\frac{4kT \, \Delta f}{R_L}} \tag{136}$$

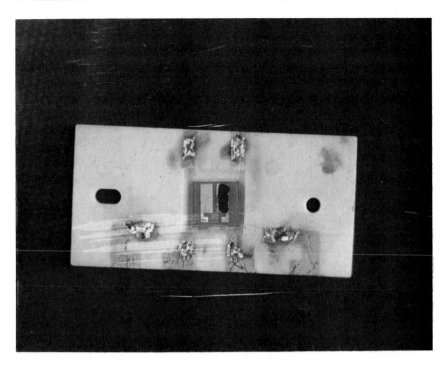

Fig. 23. Mounted and coated planar pyroelectric gas analyzer. This device uses activated charcoal as the chemically sensitive material.

If it is assumed that Δf is set by the $R_L C$ time constant, then

$$\Delta H_i\bigg|_{\min} = \frac{C_v d}{A p R_L} \sqrt{\frac{4kT}{C}} \tag{137}$$

Using the numbers in Table III in Eq. (137), $\Delta H_i|_{\min}$ is 2.4 μW. Therefore a monolayer desorption will correspond to a 10:1 signal-to-noise level.

TABLE III

CONSTANTS AND REPRESENTATIVE VALUES USED
IN CALCULATIONS

$N_0 = 10^{15}$ cm^{-2}	$A = 0.1$ cm^2
$Q_a = 1$ eV	$d = 2 \times 10^{-2}$ cm^2
$a = 2.77 \times 10^{-5}$ (s-°C)$^{-1}$	$P_0 = 2.5 \times 10^{-8}$ C/cm^2-°C
$T_0 = 300$ K	$R_L = 10^9$ ohms
$C_v = 3.2$ J/cm^3	$C = 2 \times 10^{-11}$ F

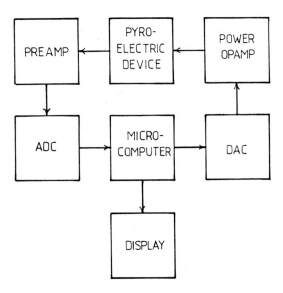

Fig. 24. Schematic of the microcomputer controller and data logger for the pyroelectric gas analyzer.

To carry out these measurements in a simple, direct fashion, it is essential to have precise control of the heating rate and measurement cycle. This was accomplished by using an eight-bit microcomputer system (Young, 1983) with the basic information flow scheme shown in Fig. 24. In essence, the software generates a steadily increasing power flow to the PGA heater electrode. This increasing power flow causes the temperature to rise at a sufficiently steady rate that a near-constant rate of increase of the chip temperature is generated. The result is a near-constant pyroelectric signal over time. When the temperature reaches the point where either the gas on the charcoal begins to desorb rapidly or a chemical reaction begins rapidly, the energy flow is modified and a large positive or negative signal is generated. The signal is measured with a 12-bit ADC controlled by the same computer that controls the DAC driving the heater. The data are stored in digital form for subsequent analysis.

4. Recent Results

An illustration of some of the data taken with the activated charcoal PGA described above is the adsorption of residual gases in a conventional oil-pumped system at a total pressure of 10^{-5} torr. This is shown in Figs. 25 and 26. The data in Fig. 25 demonstrate the signal pattern from a freshly cleaned, activated charcoal film, approximately 0.05 cm^2 in area

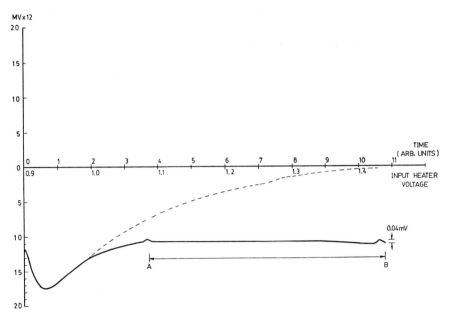

Fig. 25. Differential output voltage of the pyroelectric gas analyzer (PGA) as a function of time. The chemically sensitive layer is activated charcoal. The dashed curve represents the response to a step change in power. The solid curve is the response to a linearly ramped power input. The data were taken in an oil-pumped system at $\sim 10^{-5}$ torr after the PGA was flashed to over 200°C. Note that between A and B there is no structure. The 0.04 mV signal corresponds to a heat pulse of less than 10 μW.

and 100 μm thick. These numbers are very approximate and should be viewed as representative rather than definitive. Figure 26 shows data taken after a 3-day exposure to the vacuum. The magnitude of the peaks suggests that the gas emission was of the order of 10^{17} to 10^{18} molecules/cm^2. The positive peaks are endothermic, indicating that chemical reactions such as

$$O_2 + C \rightleftharpoons CO_2$$

$$O_2 + 2C \rightleftharpoons CO_2$$

were taking place. The negative peak is expected for gas desorption. A very interesting aspect of this work is the reproducibility of the pattern. The poor thermal contact between the activated charcoal layer and the pyroelectric vehicle limited the quantitative reproducibility in this test. Nevertheless, these data provide confirmation of the basic concept.

Another and more quantitative measure of the ability of PGA structures to monitor gases is provided by some recent work on the adsorption

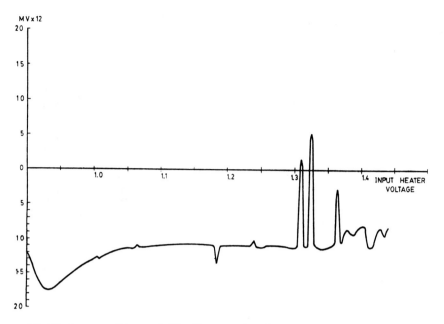

Fig. 26. Same conditions as in Fig. 25, except the PGA has remained in the 10^{-5} torr oil-pumped vacuum for 72 hr.

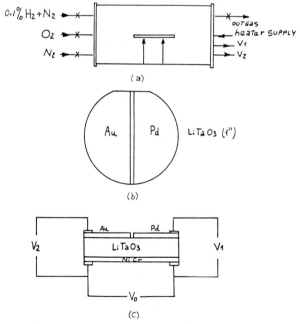

Fig. 27. Schematic layout of a hydrogen-sensitive PGA: (a) test chamber; (b) top view of the two detector electrodes; (c) circuit diagram showing electrical measurements set up on the PGA structure. (After A. D'Amico and J. Zemel, private communication, 1983.)

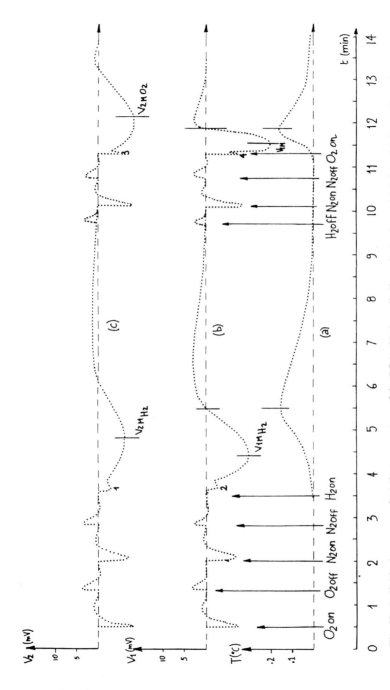

Fig. 28. Individual response of (a) the temperature of the PGA, (b) the palladium electrode, and (c) the gold electrode to the admission of various gases. The time delay between (b) and (c) is due to transport of thermal energy from the palladium electrode region to the gold electrode.

205

and reaction of H_2/O_2 and Pd thin films. In this work, a thin film of Pd was evaporated onto a single-electrode pyroelectric structure. A schematic of the PGA used is shown in Fig. 27. The Pd-PGA was exposed to a mixture of 1% hydrogen in N_2 flowing at atmospheric pressure over the device at a rate of 180 cm^3 min. When the response had saturated, the gas was changed to pure O_2 at a similar flow rate and the signal was again recorded. The resulting output signals are shown in Fig. 28. These signals are measures of the reaction between sorbed hydrogen with gaseous oxygen and adsorbed oxygen with gaseous hydrogen on the Pd surface. The data demonstrate the significantly higher energy generation rate when oxygen is admitted onto a hydrogen-saturated Pd film, compared to the rate of removal of Pd–O bonds when hydrogen is admitted to an oxygen-saturated Pd surface. Since both sorption and reaction are exothermic processes, the response curves are quite similar. The negative portion arises from the cooling of the Pd-PGA after the reactions have proceeded to completion. It should be recalled that the measured voltage is a measure of $\langle dT/dt \rangle$ and, upon cooling, $\langle dT/dt \rangle$ becomes negative. The smaller signals are due to adiabatic heating and cooling when valves are opened or closed during the gas cycles.

These preliminary results are only qualitative; however, they do indicate that there is a promising future for this type of chemical sensor based on specific chemical reactions. The theory relating the chemical reaction rates to the amount of energy derived from a particular chemically sensitive layer has been largely worked out in studies on catalysis. The specific application to these devices remains to be demonstrated.

REFERENCES

Behrndt, K. H., and Love R. W. (1982). *Vacuum* **12**, 1–9.
Bergveld, P. (1970). *IEEE Trans. Biomed. Eng.* **17**, 70.
Cady, W. G. (1969). "Piezoelectricity." Dover, New York.
Carlsaw, H. S., and Jaeger, J. C. (1959). "Conduction of Heat in Solids," 2nd ed. Oxford Univ. Press (Clarendon) London and New York.
Dalven, R. (1980). "Introduction to Applied Solid State Physics." Plenum, New York.
DeGroot, S. R., and Mazur, P. (1969). "Non-Equilibrium Thermodynamics." North-Holland Publ., Amsterdam.
Henderson, D. E., Ditaranto, M. B., Tonkin, W. G., Ahlgren, D. J., Gatenby, D. A., and Shum, T. W. (1982). *Anal. Chem.* **54**, 2067.
Herbert, J. M. (1982). "Ferroelectric Transducers and Sensors." Gordon & Breach, New York.
Hlavay, J., and Guilbault, G. G. (1980). *Anal. Chem.* **49**, 1890.
King, W. H., Jr., Camilli, C. T., and Findeis, A. F. (1968). *Anal. Chem.* **40**, 1330.
Kowalski, B. R. (1981). *Trends Anal. Chem.* **1** (3), 71–74.
Landau, L. D., and Lifschitz, E. M. (1959). "Theory of Elasticity." Pergamon, Oxford.

Lang, S. B., and Glass, A. M. (1977). "Principles and Application of Ferroelectric and Related Materials." Oxford Univ. Press, London and New York.
Lovinger, A. J. (1983). *Science* **220**, 4602.
Mason, W. P. (1950). "Piezoelectric Crystals and Their Application to Ultrasonics." Van Nostrand-Reinhold, Princeton, New Jersey.
Mason, W. P. (1966). "Crystal Physics of Interaction Processes." Academic Press, New York.
Nye, J. F. (1957). "Physical Properties of Crystals." Oxford Univ. Press, London and New York.
Putley, E. H. (1970). *In* "Semiconductors and Semimetals" (R. K. Willardson and A. C. Beer, eds.), Vol 5, p. 259. Academic Press, New York.
Smith, A. C., Janak, J. F., and Adler, R. B. (1967). "Electronic Conduction in Solid." McGraw-Hill, New York.
Stratton, J. A. (1941). "Electromagnetic Theory." McGraw-Hill, New York.
Tiffany, W. (1975). *Proc. Soc. Photo-Opt. Instrum. Eng.* **15**, 225.
Young, J. C. (1983). MS Thesis, University of Pennsylvania, Philadelphia.
Zemel, J. N. (1979). Surface Sci. **86**, 322.
Zemel, J. N., and Bergveld, P. (1981). "Chemically Sensitive Electronic Devices," Sens. and Actuators, Vol. 1. Elsevier Sequoia, Lausanne, Switzerland.
Zemel, J. N., Keramati B., Spivak, E. W., and D'Amico, A. (1981). *Sens. Actuators,* **1**, 427–474.

Bibliography

For a number of years, S. B. Lang has been reviewing the literature and publishing bibliographies at regular intervals in the journal *Ferroelectrics*. The latest of these reviews are:

"Bibliography on Piezoelectricity and Pyroelectricity of Polymers," *Ferroelectrics* **34**, 239 (1981); **45**, 283 (1982).
"Literature Guide to Pyroelectricity," *Ferroelectrics* **34**, 71 (1981); **45**, 251, (1982).

Index